Weightless

Weightless

Making Space for My Resilient
Body and Soul

Evette Dionne

An Imprint of HarperCollins*Publishers*

WEIGHTLESS. Copyright © 2022 by Evette Dionne. All rights reserved. Printed in the United States of America. No part of this book may be used or reproduced in any manner whatsoever without written permission except in the case of brief quotations embodied in critical articles and reviews. For information, address HarperCollins Publishers, 195 Broadway, New York, NY 10007.

HarperCollins books may be purchased for educational, business, or sales promotional use. For information, please email the Special Markets Department at SPsales@harpercollins.com.

Ecco® and HarperCollins® are trademarks of HarperCollins Publishers.

FIRST EDITION

Designed by Angela Boutin

Library of Congress Cataloging-in-Publication Data has been applied for.

ISBN 978-0-06-307636-5

22 23 24 25 26 LSC 10 9 8 7 6 5 4 3 2 1

For Nola, one of the lights of my life, who will hopefully inhabit a world that's far more loving than this one.

Contents

INTRODUCTION

I am in heart failure. It's a surreal sentence that still gets stuck in my mouth, coating my tongue like a dry scoop of peanut butter. I've said the sentence casually on social media and to men I'm getting to know, rebuffing all attempts at sympathy by quickly saying in one way or another, "Don't worry about me. I'll be just fine. Take good care of yourself." My boundless optimism is really a façade, a faux manifestation of the word "fearlessness," which I tattooed on my right forearm in 2014, after a particularly harrowing depressive episode that I thought would hover over me forever. Though a crinkling smile still crosses my face every day, bringing the round apples of my cheeks almost up to my eyes, and I still edit magazines, go on dates, and grapple with the angst that often accompanies turning thirty, I'm also deeply afraid for the first time in a long time.

If heart failure weren't enough, I also have stage two pulmonary hypertension, or high blood pressure in the lungs, a rare, progressive lung condition that used to kill people in two years or less. Sometimes, late at night, I comb through the #pulmonaryhypertension hashtag on Instagram. It's full of people with varying degrees of the condition carrying oxygen tanks or adjusting stomach ports that pump them with medication twenty-four hours a day. There are photos of people in the hospital awaiting double lung transplants and others trying to complete a whirlwind bucket list before they run out of time. I scroll through those pictures as if they're a crystal ball predicting my inevitable future. Sometimes it sends me into a panic attack. Sometimes it makes me sob uncontrollably. Sometimes it makes me feel grateful that the disease hasn't advanced that far . . . yet.

Fear is my normal condition. I fear that my heart will suddenly stop before I've even had the chance to fully live. I fear that I'll encounter an adverse side effect to the myriad medications I'm taking, and that it will kill me before my heart does. I fear that I'll leave behind the people who need me most—my parents, my grandparents, my brother, my aunts, uncles, and cousins, and, most important, my nieces, who still perceive me as their confidante, the one person who holds their childhood secrets close to the vest.

Now that I have less control of my body than I've ever had before and medications are the sole determinants of my longevity, writing this book makes perfect sense. I've been thinking and worrying about my fat body long before my heart failure diagnosis, so I hope my personal stories might light a pathway for other fat people to reclaim their bodies. I'm also reminded in this moment that bodies are resilient. At night, when I'm browsing that hashtag on Instagram or when I lose my breath as I ascend a staircase, I think, "My body has betrayed me," but it's not a sentiment I actually believe. It's borne from frustration, not a baked-in belief. My body has not betrayed me; it has continued rebounding against all odds. It's a body that others map their expectations on, but it has never let me down, and for this I am grateful. I need those reminders, especially in dark, quiet moments when I hesitate to go to sleep because I'm afraid that I won't wake up. Though I can't control if or when I'll recover, I can still dictate how I feel about my body and encourage others to build that level of sacred relationship with their own bodies.

I'd originally gone to the doctor because I felt an unreasonable amount of fullness in my chest—a surefire sign that I was either in the throes of an asthma attack or about to have one. I assumed the doctor would give me a professional-grade albuterol treatment, which would ease the inflammation in my lungs so that I would be able to breathe better. An asthma treatment was given, but the fullness in my chest remained. A look of concern crossed the doctor's face as she listened to my lungs and heart with her stethoscope. "You might have a blood clot that is traveling from your legs into your lung," she explained with as much calmness as she could muster. Though I felt panic in that moment—alone in a small room with bland posters on the wall—I tried to convey a mature calmness. "That's treatable, right?" I remember asking. "Is that something we can catch and cure?" The doctor nodded affirmatively, but didn't return the fake, I'm-too-blessed-to-be-stressed smile that I flashed in her direction.

When I left the doctor's office hours later, I'd taken a test that eliminated a blood clot as the reason for the chest fullness, but I did have a referral for an echocardiogram (an ultrasound of the heart) and explicit instructions to stay as calm and relaxed as possible so as not to overexert the heart. Within a week, I was gasping for air as that same doctor told me that I was in danger: my heart muscle seemed to be overworking because my heart function was considerably lower than it should be, and I needed to schedule an emergency appointment with a cardiologist. Time was as still as a frozen lake in that moment.

"You are in heart failure" is still one of the scariest sentences I've ever heard; it bowled me over in my new cardiologist's office as she explained that my heart was only working at about 16 percent of its capacity, and there was no guarantee that it would ever return to the healthy 80 to 85 percent range, even if I followed all the recommendations to the letter. I left her office with reddened eyes, two prescriptions, and a treatment plan that could turn it around or change nothing at all. The idea that my heart isn't pumping blood as well as it should pangs me, even after my cardiologist has explained the condition to me countless times, I've shared my diagnosis with friends and relatives, and I've started a rigorous regimen that limits me to 72 ounces of liquid and 2,000 milligrams of sodium per day. There's a new normal associated with my medical condition: I eat lunch at 1 p.m., a snack at 4 p.m., and dinner at 7:30 p.m. to curb the side effects of the multiple pills I take three times a day, and I have to step on a scale each morning to measure how much fluid has pooled in my body overnight. If

I gain more than three pounds in a day, then it's back to urgent care for more poking, prodding, and diagnostic tests.

The cocktail of medications I've been prescribed is designed to remove excess fluid from my body, keep the vessels in my lungs flexible and open, slow down my heart's beat and regulate it, and lower my blood pressure so no part of my cardiovascular system is overworking. My blood pressure is so low, in fact, that my body can no longer regulate its own temperature: I'm always cold, so cold that it feels as if I've been touching snow without gloves on. I also can't get too hot—super-hot bubble baths are a no-go—or I develop nausea and feel as if I'm going to pass out. My feet, legs, and fingers sometimes cramp so badly that yellow mustard—the worst condiment of all time (argue with your momma)—has permanent placement in my fridge. I need to swallow a spoonful of it every time my foot cramps become unbearable. I drink a glass of orange juice every morning, take two prescription-strength potassium pills every afternoon and a B_{12} pill at night, and eat at least one banana every day, and it's still not enough to ward off the cramping. These are just the prices my body pays for being resilient.

Heart failure has stolen a lot from me: Walking a block makes me feel as if I've run a marathon. My fingers swell so badly that I sometimes can't wear the glitzy rings that have long been staples in my wardrobe. And I spend hours in doctors' offices, having blood drawn, taking diagnostic tests, and figuring out if my medicines are working or making my heart—and possibly my kidneys—worse. I can't work as much as I used to. I can't even enjoy my typical meal at Chipotle, which has too much salt for my sodium restrictions. (This is the real tragedy of it all.) The very energy that once convinced me that I could write two books, work full-time, and still sleep eight hours each night is gone. There are days when it is difficult to muster the strength to get out of bed, brush my teeth, and get dressed. But as much as these chronic conditions have drained me, they have also helped me reframe what's important—advocating for myself and for my needs with doctors, romantic partners, bosses, coworkers, and every person and system I interact with. Heart failure has also reminded me that I'm not alone in this fight to assert my agency. I've never been in it alone.

My parents have always been a salvation for me. They realized early on that I was bigger than my classmates, but they didn't panic, force me into dieting, or give me a sense, as in other autobiographical fat girl stories, that I was undeserving of starting my life until I was smaller. Of course, there were boys who saw a girl who'd gotten her first period at ten and thought I was adult enough to be groped and subjected to sexist mistreatment. There were snide comments from fellow kids about my fat cheeks. But my parents consistently, incessantly, and without apology told me that I was worthy. When my teachers and classmates bombarded me with fatphobia, I felt their love for me in the comfortable confines of our three-bedroom apartment. It came through my mother wrapping me in her arm as we sat on the couch, rubbing my head in a rhythm from front to back, and assuring me without saying a word. It was a bubble of love—their attempt to arm me with the protection I needed to navigate a world hell-bent on my destruction.

My mother, who has a complicated relationship with her own plus-size body, understood better than anyone what it's like to navigate a thin-obsessed world. Though she's embarked on diet after diet, she tried to prevent me from falling into a dieting spiral. My father seemingly knew he'd be unable to protect me after I'd left our family's sky-blue station wagon every morning, so he used our twenty-minute morning drives to try to create an invisible force field around me, an impenetrable shell that would shelter me from the cruelty that's as common among children as chicken pox. As he drove me toward the taunting that surely awaited me, tired after working a twelve-hour overnight shift, he'd allow me to read to him. It didn't matter what we read—he would let me choose without pushback. Sometimes, it was a Nancy Drew or Hardy Boys mystery; other times it was the latest edition of The Baby-Sitters Club or Sweet Valley High.

He always listened attentively, never once revealing how tired he was. After reading to him for fifteen minutes, I'd pepper him with questions to ensure he'd been paying attention. I was nothing if not a schoolteacher in the making: "What was the villain's name? What was the biggest lesson you learned?" He could've (and probably should've) told me to be quieter, but instead, he was attentive, answering every question and reminding me to use a bookmark so we could pick up where we left off.

Beyond indulging me, he was also inadvertently giving me access to my voice. In his presence and under his guiding hand, I wasn't subjected to the whims of a world obsessed with thinness. I was powerful. What I read and said meant something. My father unknowingly tapped into something important that has guided me through fatphobic, racist, and sexist discrimination, harassment, mental illness, college, grad school, and now my career: no matter the size of my body, nothing—including heart failure—can hinder me from living a full and complete life. Today, my parents' quiet and calming presence at doctors' offices, in the home we share, and hovering above me as I lie in bed, trying desperately to gather the energy to start the day, gives me an unexplainable peace. No matter what comes next on my health journey, I know I have the support I need to survive. Don't we all deserve that? We all deserve communities that show up for us, pick us up when we're falling, and stop for us when we have to catch our breath.

In school, I wasn't as protected as I tended to be at home. I was treated as an object to be surveilled, controlled, touched without consent, and rendered helpless. Though I had a close group of friends with whom I ate lunch with every day and had sleepovers, camaraderie was no match for the harassment I endured day in and day out. By the time I realized my body made me different, and somehow more vulnerable, than my elementary school classmates, it was too late. It happened in fifth grade when my teacher asked me to take the class trash to the dumpster and asked two of my peers to accompany me. Two classmates, both Black boys who prided themselves on slapping girls on their asses without their permission, cornered me a few feet away from the green dumpster into which we would toss the garbage. One of the boys, brown, with the curliest ponytail I'd ever seen, made a joke that I've long forgotten. The other, a chocolate kid with a fade that he never seemed to brush, was laughing one moment and pushing me down the next.

He pinned me to the concrete, then climbed on top of me, wrapping my hands in his so I couldn't escape. He looked down at me with an emotionless glare that still causes the hair on my arms to rise when I remember it. I'd never felt more afraid. As he breathed heavily, he simulated sex, rubbing his genitals against me through the jeans my father had picked out for me to wear that morning.

It lasted for two minutes or maybe twenty or maybe an hour or maybe forever, but he quickly got up when I yelled for help, dusted off his pants, and reached out his hand to help me up. He laughed, as if this were just another version of the ass-grabbing game he'd initiated on the playground. Then he walked me back into school, his arm around my shoulder, comforting me and, through the pressure he applied, reminding me that this was our secret. We resumed our friendship, and I never brought it up—to him, my parents, a teacher, or anyone who could have intervened. I was ten.

As a result of this experience, I had a dawning realization: no matter how I felt about my body or the positive language my parents used to affirm my widening hips and budding breasts, my plus-size body didn't belong to me.

By the time I started junior high school, I had a very clear understanding that my body was a public spectacle because I had C-cup breasts, hips, a menstrual cycle, and a wide ass to match. I had to pay a price for the literal amount of space I took up in the world, and the penalties rained down: There were the untoward comments about my weight and the size of my breasts ("Here comes Evette with the big C cups" was a taunt that followed me from fifth grade through middle school). There was the unwanted touching from classmates, including a seventh grader who insisted on rubbing my thigh in science class and another male student who grabbed my breast as I passed him in the hallway.

And, of course, there was the gaslighting: Teachers turned the other cheek when boys taunted me in the hallway. When my father witnessed a male student smack me on the behind, causing me to drop the gingerbread house I'd spent all afternoon working on, he confronted the boy and the boys' parents, who apologized for their son's behavior and made him apologize for touching me without my permission. Yet, when my father brought his concern to the school's administration, they brushed him off. How could they control harassment? And more than that, these were just preteen antics—something women the world over experienced as children, a training ground where we're taught to accept mistreatment with impunity. Since we live in a fatphobic culture that's bolstered by a billion-dollar dieting industry, desire is supposedly reserved for those who fall within the constraints of the thin ideal. So, all the attention that plus-size girls receive, whether it's wanted or not, should flatter us—a lesson that some people learn from how fat women are depicted on-screen.

After being pinned to the ground and harassed, penalized for being fat, and blinded by diet culture, I realized that many people believe that fat girls are undeserving of fairy tales, of respect, of the invisibility afforded to bodies that are deemed "normal." When I began writing this book, I created very clear boundaries about what this collection of essays would be and what it could never be. *Weightless* isn't strictly a memoir, though it includes personal stories that illustrate my experience of moving through the world in a body that's often treated as if it's repulsive. This isn't a book about dieting. I'm not writing about a lifelong quest to lose weight. I'm not writing about an inherent hatred that I'm unable to quell because I've chosen to remain in a fat body. I'm not writing a book about fatness that's inherently about a personal journey to wholeness—number on the scale be damned.

Instead, *Weightless* is an excavation of a culture that hates fat people and uses institutions, including media, medicine, and marriage, to reinforce that repulsion. Whether it's Netflix greenlighting a television show that glorifies losing weight as a form of revenge or airlines enacting policies that purposefully discriminate against fat people, the world believes that we must assimilate and become smaller—not that it should become bigger to accommodate us.

Weightless doesn't shy away from discomforting topics, because to exist in a fat body is to be made to feel uncomfortable all the time. It's a book designed to shift how we individually and collectively understand the fat experience. Most important, it's a book for fellow fat people who want, need, and deserve new stories. Fat people aren't a problem that needs to be solved. Fatphobia, which creates a world in which we're all made to believe that thin bodies are better and deserving of better treatment, is the issue—and that's where my focus lies. In this book, I interrogate not only the fatphobia that we exhibit toward one another, but also how it is reflected and reinforced through the pop culture we consume.

Fatphobia is so persistent in American culture, so invasive, and so sneaky that it has become an everyday part of our lives. It intertwines itself in our language—in how we talk about our own bodies and other people's bodies. But it doesn't have to be this way. Body positivity, a movement to dismantle systems that target, shame, and map stigma onto fat bodies, is having a cultural moment. Women of size are staking our claim to liberation and happiness through fashion lines, bikinis, and magazine covers. We're declaring our right to exist as we are without persecution, whether that's through a fatkini revolution or *Refinery29*'s 67% Project, which aimed to make the 67 percent of American women who are plus-size more visible through the pictures their editors chose for stories and an empowering set of photos launched with Getty Images. Fat women are investing in and gravitating toward images of ourselves, and the impact is felt: plus-size fashion collections sell out as soon as they're released, while Iskra Lawrence, Ashley Graham, Gabi Gregg, and a number of other plus-size models, fashion designers, and bloggers have been elevated to icon status.

Despite fashion and the focus on fat liberation in women's media, pop culture, including television and movies, continues to offer a one-dimensional view of fat women and we're still mistreated in our everyday lives. On-screen, we're asexual, comedic relief. We don't have full lives or partners who value us. We have sex with the lights off. We're obsessed with our weight. In doctors' offices and on airplanes, we're subjected to fat-shaming, which makes it even more difficult to move through the world. So, how can we survive in the real world? After being diagnosed with heart failure and pulmonary hypertension, conditions that weren't caused by my size but were overlooked because of it, I know now what fatphobia costs fat people: it robs us of our joy, our ability to progress in our careers, and, sometimes, in cases all too similar to mine, our lives. Fatness isn't a choice, but as long as it's treated as if it is, we'll have to do dual work: steeling ourselves against discrimination while working to dismantle fatphobic systems, from the media to medicine.

I started writing *Weightless* before I became ill and I am finishing it after a life-altering diagnosis, and in the process, I've learned new lessons about how fat people are regularly dismissed, willfully misinterpreted, and sidelined as they advocate for their own health and well-being. *Weightless* offers a three-dimensional view of the fat woman through the lens of a fat woman who's found that, in some ways, self-love is a greater force than fatphobia. Let's begin.

Weightless

NO COUNTRY FOR FAT KIDS

In 2018, Weight Watchers announced that they would begin allowing teens between the ages of thirteen and seventeen to join for six weeks for free. They did this for two stated reasons: to increase sales from $1.2 billion in 2016 to $2 billion by the end of 2020 and to start building a younger, loyal fan base. *CNN Business* reported that sales jumped 16 percent when Weight Watchers—rebranded as WW—announced the initiative, and Oprah Winfrey, who owns a 10 percent stake in the company, said that she was "inspired to be part of this purpose-driven mission" that makes "wellness accessible to everyone."

Yet, offering teenagers free access to dieting has unintended consequences, as reporter Anna Medaris Miller found when she interviewed teens for *U.S. News & World Report* about WW's announcement. Becca, who asked Miller not to use her last name, began the Weight Watchers program when she was twelve, and it severely damaged her self-esteem. "It was instilled in me that something was wrong with me and that I was wrong because of the way I looked, and so I tried to change the way I looked," she said. From age twelve to twenty-two, Becca spun into a cycle of dieting and food obsession that slowly devolved into an eating disorder. "It's taken me a lot of years of therapy to not see food in points, and see it as food," she told Miller. Unfortunately, Becca's experience is common, as registered dietician Rebecca Scritchfield explained. "We have no long-term data that shows that trying to suppress kids' weight is successful at long-term weight suppression and improves their well-being," she told Miller.

I know the effects of Weight Watchers all too well: I joined the program in 2013 after getting to my largest weight in graduate school. Poor eating habits combined with depression and school-related isolation led to a fifty-pound weight gain in less than eight months. I ate excessively as I studied, wrote my thesis, and taught classes to undergraduate students. Pummeling through multiple plates of food helped me suppress the depression I'd slowly begun to sink into. One afternoon in August, a mere three weeks after my second year of graduate school began, I realized my favorite pair of jeans no longer fit. I immediately pulled out the scale I'd buried at the bottom of a storage container. The number blared at me as I looked down at my weight. I joined Weight Watchers that night.

I'd witnessed my mom use Weight Watchers to lose a significant amount of weight before gaining it all back, and I figured that I could replicate her initial success. I did. I lost a significant amount of weight over the course of four months, thanks to the program's point system. Weight Watchers assigns a specific number of daily points to dieters based on their height, weight, and goals, and each food item has a different point value. Weight Watchers sells their own snacks, frozen meals, and recipe books to assist customers. They also have weekly weigh-in meetings to hold dieters accountable to their goals. I quickly fit back into those jeans I loved, thanks to Weight Watchers, a daily intake of 80 ounces of water, and a personal trainer. However, Weight Watchers also had an unintended impact: for the first time, I became obsessed with food and dieting.

I dreaded Weight Watchers meetings. I always made sure I wore as few clothes as possible, drowned myself with water in the morning so I could urinate more throughout the day, and ate as little as possible. Weighing less than I did the week before became an obsessive game for me. I worked out harder to burn more calories. I avoided "heavier" foods, like pastas and sandwiches, as I neared a weigh-in day. While Weight Watchers prides itself on blending seamlessly into any dieter's life, it doesn't account for the ways that we have to reshape our lives to consider the program. I planned my days around points. Could I eat at a restaurant without going over my points? Would I extinguish my points by having a slice of cake and a sandwich within the same day? I began seeing a therapist for depression about six months into the program. Many of my early sessions were spent lamenting about shame. I felt shame each time I hit a weight plateau or ate more than my points allotted for, which only fed into my depression. My counselor helped me process through the way I used food to bury emotions—and encouraged me to quit Weight Watchers.

Anything that made me feel so poorly about my body wasn't mentally or emotionally healthy. I quit Weight Watchers after eight months—and felt freer than I did the entire time I adhered to it. This depressive cycle of losing, plateauing, and gaining destroyed me emotionally, and I was an adult. Imagine what WW's program—and those built within its image—can do to children who are less aware of the impact of dieting culture on our mental and emotional health and are therefore more susceptible to seeing their bodies as flawed and in need of fixing. For many, using WW—or, let's say, a program ushered in by a presidential administration—could be a gateway to disordered eating.

On March 20, 2009, First Lady Michelle Obama and twenty-three fifth graders from Bancroft Elementary School in Washington, D.C., broke new ground—literally. They began digging up a 1,100-square-foot plot of soil on the South Lawn to plant the White House's second-ever vegetable garden. The last First Lady to embark on such an endeavor was Eleanor Roosevelt, who planted a "victory garden" to encourage American families to grow their own fruits and vegetables. By May 1943, eighteen million victory gardens were producing more than one-third of America's vegetables. Gardening, as it was framed, was a patriotic endeavor, a fight for the future of the United States. "Food Will Win the War" was a popular phrase that appeared on ads and posters throughout the United States.

The idea that homegrown vegetables are essential to ensuring the future of America—and in this case, America's children—has continued through to today. The First Lady wanted to put the vegetables from the garden on the plates of President Obama and their daughters, Malia and Sasha, and serve them during state dinners. She told NPR in 2012 that honey from the garden's beehive was even used in her tea and the White House's home-brewed beer. However, feeding the First Family wasn't Obama's only motivation for planting the garden; it was also her first step toward educating children and their families about the importance of including fruits and vegetables in their daily diets.

"My hope is that through children, they will begin to educate their families and that will, in turn, begin to educate our communities," she told *The Washington Post* the day before she began planting. "I wanted to be able to bring what I learned to a broader base of people. And what better way to do it than to plant a vegetable garden in the South Lawn of the White House?"

There was no better way for Obama to begin her mission to address what she perceived as one of America's biggest issues: childhood obesity. As U.S. surgeon general Richard Carmona explained at the 2004 Summit on Obesity sponsored by *Time* and ABC News, "Childhood obesity . . . is every bit as threatening to us as is the terrorist threat we face today. It is the threat from within." Obama argued that America's children would eat better if they were involved in the process of growing and preparing their own food. "My girls like vegetables more if they taste good," Obama told *The Guardian* in a 2009 interview. "Especially if they're involved in planting it and picking it. They were willing to give it a try." The implication here was that giving children access to fresh fruits and vegetables would have a ripple effect: children would eat healthier, become more active, and, in the process, lose weight—without developing unhealthy relationships with their bodies or a lifelong obsession with being thin.

On February 9, 2010, Obama took this idea a step further by announcing her first initiative, Let's Move!, an ambitious, multipronged program that sought to reduce childhood obesity by 5 percent by the year 2030. The initiative would work to change nutrition policies as well as encourage children to become more active. With assistance from a number of organizations and colleges, including Kaiser Permanente, the Robert Wood Johnson Foundation, Anthem Blue Cross Blue Shield, Aquafina, Boys & Girls Clubs of America, Nike, and Oregon Health & Science University, Obama created the Partnership for a Healthier America to provide private sector resources for three key facets of Let's Move!: creating healthier places, increasing physical activity, and transforming the marketplace.

That same day, Barack Obama signed a presidential memorandum to initiate the Childhood Obesity Task Force, a partnership among multiple federal agencies, including the Office of the First Lady, the Department of the Interior, the Department of Agriculture (USDA), and the Department of Health and Human Services. While Mrs. Obama focused on public messaging, including tapping celebrities like Beyoncé and Shaquille O'Neal to champion the initiative, the Childhood

Obesity Task Force focused on crafting policies to support her mission. Within three months, the task force offered four recommendations designed to end childhood obesity within a generation: (1) empower parents and caregivers to make healthy choices for their families; (2) serve healthier food in schools; (3) ensure access to healthy, affordable food; and (4) increase physical activity.

Within a year of Let's Move! being announced, Congress passed the Healthy, Hunger-Free Kids Act of 2010 to increase access to healthy food among low-income children. The reauthorized bill guaranteed funding for federal school meals and childhood nutrition programs for five years and included $4.5 billion to fund the National School Lunch and Breakfast programs; the Special Supplemental Nutrition Program for Women, Infants, and Children; the Child and Adult Care Food Program; the Summer Food Service Program; the Afterschool Meal Program; and the Supplemental Nutrition Assistance Program Education for ten years. The act also included controversial provisions, such as allowing the USDA to set nutritional standards for foods sold in schools, including in vending machines and in school stores; enforcing nutritional and physical activity goals at schools; and requiring schools to submit to audits every three years to ensure their compliance with the USDA-set nutritional standards.

Outside of schools, Let's Move! and the Childhood Obesity Task Force worked with the Food and Drug Administration to make food and beverage labels friendlier and easier for families to read. They also partnered with the American

Academy of Pediatrics to ensure that pediatricians tested all children's body mass index (BMI) during wellness visits to gauge if they were at risk for developing obesity. Media companies, including Disney, NBC, and Viacom, agreed to air Let's Move! public service announcements and collaborated with the Obamas on shaping their public awareness campaign.

On the surface, the White House's mission seemed well intentioned and rooted in Michelle and Barack Obama's concerns about their own daughters. In 2010, Michelle Obama said that her family's pediatrician told her that Sasha and Malia were gaining weight too quickly because she was choosing to feed them pizza and fast food rather than fresh fruits and vegetables. "In my eyes, I thought my children were perfect," Obama said during an appearance at a YMCA in Alexandria, Virginia. "I didn't see the changes." After her doctor's warning, Obama made incremental changes to her daughters' diet as well as their afterschool activities. For instance, she stopped allowing Sasha and Malia to watch television during the school week; the American Academy of Pediatrics found that 25 percent of American children watch at least four hours of television per day, and those who watch more than that are more likely to have a higher BMI. Obama also fed Sasha and Malia smaller portions, put water bottles in their lunch boxes, served them only low-fat milk, and encouraged them to eat healthier by giving them fruits and vegetables at every meal.

Let's Move! seemingly made sense in a country where nearly one-third of kids are overweight and 17 percent are obese, but Let's Move! and the Childhood Obesity Task

Force didn't consider one key facet: how difficult it would be to help fat children gain and maintain ownership over their bodies in a culture obsessed with thinness. Let's Move! and the Childhood Obesity Task Force didn't provide appropriate resources to the families of children who are taught that their bodies don't belong to them, and that the price they must pay for being larger is constant and unrelenting surveillance. And, most important, Let's Move! and the Childhood Obesity Task Force didn't account for how America's history of prioritizing political and economic interests at the expense of marginalized communities contributed to the nation's "childhood obesity crisis."

I understand far too well the uninhabitable world that this focus on childhood obesity creates for fat kids. I lost control of my body when I was eight. Until then, doctors had been kind, never once expressing concern about the size of my gangly arms, pencil-thin legs, or flat chest. My teachers talked to my parents about how I was performing academically rather than my ability to climb up a rope in gym class. The world seemed to bend to my childhood whims when I fit within an acceptable BMI. When I went to the park, running with abandon as the crisp fall wind whipped against my cheeks, I wasn't concerned about fitting on the slide or being too heavy to be pushed in a swing. I was so carefree then.

Everything changed when I suffered two consecutive asthma attacks that landed me in the hospital for days. I first felt my lungs constrict as I was waiting for my father to pick me up after school. I was on the concrete bench adjacent to the school, where students were expected to sit quietly until our parents arrived. It started as nausea. Suddenly, a roiling pain unlike anything I'd ever felt before—resembling the menstrual cramps that would strike me less than two years later—overtook me. It felt like I'd eaten something rotten that was trying to force its way out of my stomach. Then, the wheezing began. I struggled to catch my breath as I heaved and heaved until my stomach emptied and a teacher stood over me and rubbed my back.

By the time my father pulled up five minutes later, I'd recovered, caught my breath, and was again sitting quietly. By the end of the week, my parents were rushing me to the emergency room as I struggled through ragged breaths, hoping that one would break up the phlegm in my chest and allow me to take a full breath. At first, my doctors didn't diagnose me with asthma; instead, they sent my parents and me home with a nebulizer to use whenever I felt congested, antibiotics to cure the infection looming in my lungs, and paperwork that explained the ins and outs of bronchitis. The doctors' diagnosis and treatment plan seemed to work until a month later, when an intense bout of wheezing signaled that I wasn't getting better. Overnight, my lips turned a deep shade of blue.

This time, I spent seven days in the hospital, hooked to a machine that pumped me with Prednisone, a steroid that would strengthen my lungs, ward off asthma, and allow me to leave the hospital alive. The doctors filled me with so much Prednisone and albuterol, a common medication used to treat asthma, that my hands shook as I attempted to eat the hospital-prepared lunch. What the doctors didn't tell my parents—or me—was that Prednisone would transform my body and send me into a spiral of body shame that would take years to unlearn.

Prednisone, which I took every morning for three months, made my face fuller—as if I'd stuffed walnuts in my cheeks. Within six months, I'd sprouted breasts, completely skipping the sports bra phase, and I'd watched my mother silently cry, tears pooling on her chin, as she helped me select bras to try on in a Sears dressing room. Suddenly, my body was too large for the pediatrician's charts. I had gained thirty pounds. At each appointment, the doctor would measure my breathing capacity and then talk quietly to my parents about putting me on a diet or appetite suppressant to curb my weight gain. In those moments, as nurses and doctors discussed my body without directly talking to me, I recognized the stinging pain of isolation. I'd sit on the exam table, absently swinging my legs and pretending to be absorbed in my childhood thoughts. Then, when we were free to go, I would slide off the table and into my coat, smile at the doctor, and thank him for his assistance. I curled into myself, but after all, I'd asked for the maligning. I was going to the corner store for a Slurpee and Kit Kat every day after school, and, surely, that was contributing to my body expanding at such an alarming rate.

While Let's Move! and the Obesity Task Force discuss childhood obesity in the abstract, the reality is that fat children are bombarded with reminder after reminder that their bodies are abnormal and need to be fixed. In a 2010 study published in the journal *Pediatrics*, researchers followed eight hundred children born in 1991 who lived in ten cities across the United States. Julie Lumeng, a professor of pediatrics at the University of Michigan, and her team of researchers followed those children through third, fifth, and sixth grades to determine if there was a correlation between their BMI and how often they were bullied. The findings weren't surprising: "obese" children were 65 percent more likely to be bullied than normal-size children, while overweight children were 13 percent more likely to be bullied.

A 2015 study conducted by researchers at the Rudd Center for Food Policy & Health at the University of Connecticut in Hartford and published in *Pediatric Obesity* reached a similar conclusion. Nearly 70 percent of the participants, children based in the United States, Canada, Iceland, and Australia, said that weight-based bullying is a "serious" or "very serious" problem.

Even overweight and obese children with good social skills are subjected to bullying because, from a young age, children are taught to associate thinness with discipline while fatness is treated as a sign of laziness. My own nieces, who are coming into their preteen adolescence, have told me that they don't want to be fat, and they choose to eat grapes instead of chips to keep their bodies as small as possible. They casually mention "eating clean" like it's another game of Roblox and have teetered between eating as many Takis as they can get their hands on and munching only on fruit. Bullies can sense other children's vulnerabilities, and since we live in a thin-obsessed culture, fatness is immediately pinpointed as a difference that can be fuel for torment.

Couple that with the American Medical Association's recommendation that schools teach students from kindergarten onward about the "dangers of obesity," and we have an entire generation of children with a target on their backs. When schools surveil fat children, and pediatricians surveil the BMI of their child patients to try to persuade their families to participate in interventions, it becomes even clearer how fat children are consistently targeted.

These stakeholders have historically failed to account for how policies around government assistance and school lunches have made it difficult for families from marginalized communities to provide "healthy" meals—a point that the Childhood Obesity Task Force did attempt to address—which further contributes to the rapid increase of "childhood obesity" in the United States.

In 1900, thirty-four states required children under the age of fourteen to attend school, but there were no laws that guaranteed students would receive adequate nutrition throughout the school day. Four years later, sociologist Robert Hunter sounded the alarm on the effect that an economic recession was having on almost every American child's ability to eat and focus in school in his treatise *Poverty*:

> The poverty of any family is likely to be most serious at the very time when the children most need nurture, when they are most dependent, and when they are obtaining the only education which they are ever to receive. Guidance and supervision of the parents are impossible because they must work; the nurture is

insufficient because there are too many hungry mouths to feed; learning is difficult because hungry stomachs and languid bodies and thin blood are not able to feed the brain. The lack of learning among so many poor children is certainly due, to an important extent, to this cause.

In response, Philadelphia, Boston, Milwaukee, Chicago, and a few other American cities partnered with local charities to create affordable school lunch programs. In Philadelphia, the Starr Centre Association began serving hot penny lunches in nine schools. Cheesman A. Herrick, the principal of the William Penn High School for Girls, petitioned the Philadelphia School Board to establish nutritional guidelines for school lunches and to place a home economics graduate in charge of his school's lunch program. Other cities followed suit, and school lunch programs became ever more popular among American families, who knew that their children would get at least one hot meal per day, even if they couldn't afford to feed them at home.

When the United States began staring down the Great Depression in the 1930s, the federal government intervened by buying surplus crops from down-on-their-luck farmers to feed to students and creating jobs in school cafeterias for women who were struggling to find work outside of the home. *Time* notes that by 1941, "federally supported school meals programs were operating in all States, plus the District of Columbia and Puerto Rico, with 64,298 individuals serving over 2 million lunches daily." And in 1946, as America recovered from its involvement in World War II, Congress decided to pass the National School Lunch Act to "safeguard the health and well-being of the Nation's children."

The National School Lunch Act included mandatory nutritional requirements determined by the secretary of agriculture, provided reduced-cost meals to low-income children, and allocated at least $10 million to assist school districts across the country with purchasing equipment for the program. Then came Ronald Reagan's cheap ass. In the '80s, Reagan declared war against excess spending and government waste, and instead of cutting from the military budget or the salaries of congressional leaders, he decided to slash nearly $1.5 billion from the $5 billion earmarked for school lunches and other childhood nutritional programs. That same president also declared that ketchup is a vegetable so that it could be counted as a form of nutrition in school meals. This led Republican Senator John Heinz, whose family owned Heinz ketchup, to say, "Ketchup is a condiment. This is one of the most ridiculous regulations I ever heard of, and I suppose I need not add that I know something about ketchup and relish—or did at one time."

Though the USDA rescinded Reagan's ridiculous recommendation one month later, Reagan's cuts forced schools to get into business with corporate contractors who agreed to lower costs in exchange for allowing them to dictate the schools' lunch menus. At the time, Mary C. Jarratt, the assistant secretary of agriculture for food and consumer services, told *The Christian Science Monitor* that there would be generous upsides to allowing schools to choose what they served students, including smaller portions and additional nutrition options. "The proposed changes are designed to maintain the nutritional balance of the oldest federal feeding program in the country, while at the same time providing state and local officials more flexibility in controlling costs and simplifying administration," she claimed.

What the Reagan administration never explained, however, was how nutritional standards became less of a concern than costs. The food pyramid that schools used to rely on was tossed aside in favor of name-brand pizza and soda and vending machines stocked by multinational brands, while large-scale catering companies provided bland meals like assembly-made cheeseburgers and french fries. At the same time, the USDA allowed schools to increase the price of school lunch while also tightening the eligibility requirements for reduced-cost lunches. Within months of Reagan taking office, the price of a reduced-cost meal had increased by $0.40 and an unsubsidized meal by $1.30. With less money and the same number of children, schools began cutting corners at the expense of the health of students—going from small portions of meats and vegetables to spreads of packaged desserts and french fries.

In 2009, the average school lunch contained 1,375 milligrams of sodium, more than half of the sodium that a child was supposed to consume in a single day. So, does anyone find it surprising that America's children are continuing to gain weight at what's perceived as an alarming rate? Or that Michelle Obama received so much pushback when she tried to get multinational corporations out of America's schools? Privatizing school lunches, labeling condiments as vegetables, and then increasing the cost of said lunches was cause for catastrophe. But somehow, children and their families are now to blame for being larger—not the politicians who were responsible for overseeing school nutrition policies. Talk about irony.

Five years after Let's Move! began, the overall rate of obesity was still three times higher than it had been in 1974, though the Centers for Disease Control and Prevention claimed that obesity dropped 43 percent among children between the ages of two and four. A 2018 study published in *Pediatrics* found that there is no evidence of a decline in obesity in any age group. "Despite intense focus on reducing the U.S. childhood obesity epidemic over the past two decades, our progress remains unclear," the researchers wrote. When fat children are told in every aspect of their lives that they're not enough—from school to the doctor's office to the internet—it further ostracizes an already vulnerable population.

I can picture his face so clearly: though I can't remember my elementary school teacher's name, I can see his tall, lanky frame, red glasses, and small salt-and-pepper-speckled pompadour. In his triweekly PE class, he focused his attention on students like me, using his booming voice to encourage those of us who were larger than other students to run harder and faster so we could burn more calories. "Evette!" he'd scream as I lagged behind the rest of the class while we jogged four laps around the gym. "Pick up the pace! You can keep up. Stop being lazy. You can do it!" Though he harassed me on a regular basis, there is one particular memory that still haunts me. I was barely jogging since my breasts had grown far larger than my classmates', and I was self-conscious about them bouncing and bringing more attention to me. I was also worried about having an asthma attack, a constant fear that determined how active I was in school and outside of it. "Leave me alone," I muttered under my breath as I pretended to move my feet faster.

"Go! Go! Go! Go!" he screamed at me over and over again. Once I'd finished my final lap, with all of my classmates' eyes on me, I went to sit beside them, but my PE teacher had other plans. Before I could sit, he told me to run another lap. My jaw dropped in disbelief. I'd barely managed to finish the first mile, and he was standing in front of me, preventing me from sitting down, and forcing me into an unwinnable situation: either I'd be humiliated by running a lap while all of my classmates watched or I'd be kicked out of class. I chose to run.

To their credit, my classmates didn't laugh, but they did stare as I ran as quickly as I could around the gym. It seemed like that lap took forever, as my peers' eyes bored into my back. I held back tears, sniffling and trying to keep my breaths even, as I ran toward the finish line. Once I'd finished, sat down beside my friend, and wiped my wet eyes with my shirt, the PE teacher bent toward me. "It's good for you," he said softly. "Exercise is good for you."

In that moment, I realized that my body, which was getting larger by the day, was unacceptable. It was so unacceptable, in fact, that I was being singled out in gym class to try to correct the problem. It wasn't the last time that this teacher targeted me because of my size. Whenever he monitored recess, he forced me to participate in one of the impromptu games of kickball, tetherball, or hopscotch that other students started. He always tried to frame it positively, as if he were using his influence to perform a public good, but all I heard was that my body needed to change—immediately. This level of surveillance doesn't persuade fat children to lose weight. Instead, it simply underpins the idea that we have a flaw that must be fixed. On days when someone either harassed me or pretended as if they didn't see another student or teacher targeting me, school felt unbearably difficult. When the bell rang, I'd find my way to a convenience store and stuff my face with snacks until I felt numb. "Eating your feelings" is a (rightfully) fraught term, but it applied to me during that time; I drowned my hurt in ice cream and Twix and Cheetos because I didn't feel as if anybody would really hear me if I screamed, cried, or expressed my emotions in any tangible way. Food became a form of solace as I stared down cruelty.

Even the public health campaigns meant to prevent obesity can contribute to the stigma, researchers say, because the implicit message is that anyone who really wants to—anyone who eats well and exercises regularly—can be thin. And that's simply untrue. "As weight loss programs, diets don't work! Yes, you lose weight, but about 95 percent of people who lose weight by dieting will regain it in one to five years," former school counselor Meg Selig explained in a 2010 article for *Psychology Today*. "Since dieting, by definition, is a temporary food plan, it won't work in the long run. Moreover, the deprivation of restrictive diets may lead to a diet-overeat or diet-binge cycle. And since your body doesn't want you to starve, it responds to overly-restrictive diets by slowing your metabolism, which of course makes it harder to lose weight."

In other words, forcing children to run more, consume fewer carbs, and avoid eating that doughnut after soccer practice probably won't help them lose weight. It might send them into a tailspin of lifelong restrictive dieting, but it won't make them smaller or happier. Childhood obesity rates have remained constant because nobody is invested in helping children understand their bodies, teaching them to listen to their hunger cues, or introducing them to physical activities that they might enjoy. Instead of telling children that they "don't need ice cream," we could ask them why they want ice cream. Are they feeling sad? Are they bored? Is there something comforting about eating ice cream? Instead of telling children to run an extra lap or making them feel bad because they'd rather read at recess than participate in an impromptu game of kickball, why not ask them what kinds of activities they like? Why not figure out if they're being bullied or if they're being made to feel unwelcome instead of assuming that they're just lazy?

By making nutritional choices for children based on our expectations of what their bodies should look like, we're only reinforcing a culture that values thinness at all costs. Was that Michelle Obama's intention? No. In fact, reducing childhood obesity by 5 percent could be considered a noble goal for a mother concerned about the nation's children, including her own. But Let's Move!, Weight Watchers' free membership for teens, and other programs that hyperfocus on the size of children's bodies do more harm than good. Study after study has shown that dieting doesn't work; focusing on what fat children eat rather than on the reasons they eat and how their palette is chosen will also create another generation of children with unhealthy relationships with their bodies. That makes vulnerable kids even more vulnerable. And nobody needs that extra target on their back.

DOCTORS, GET YOUR SHIT TOGETHER

If you run through the lobby again, you'll have to go home for the rest of the day," the afterschool counselor calmly told me as I slowed down in the expansive lobby of our neighborhood's local recreation center. "You could get hurt, or you could hurt someone else." I nodded in agreement—signaling that I'd heard her directive—as I tried to calm the electricity coursing through my veins. For the next thirty minutes, I walked quickly from the gymnasium to the arts-and-crafts room and back again, trying my hardest not to run—until someone called my name, and I sprinted through the lobby toward the gym. Before I could speed past her, the counselor grabbed my forearm and turned my body to face her. "That's your second warning, so you have to leave for the day," she said, with as much gentleness as she could muster. "You can return tomorrow, and I hope that you'll follow the rules then."

I was eleven at the time, and I'd endured far worse than

being chided for breaking the rules at a recreation center. A few weeks before, I'd been harassed in a 7-Eleven by an adult stranger, who tried to force me to exchange my Kit Kat for a granola bar. But that moment in that recreation center broke something deep in the well of my heart and mind that sent me into a downward spiral of depression, anxiety, and, eventually, agoraphobia. Until that point, the recreation center had been a foundational part of my childhood. When my parents decided to move our family from Queens, New York, to Denver, there was a lot of uncertainty about childcare. We would be thousands of miles away from the caretaker who had looked after us from the time my brother was six months old, and we wouldn't have the kind of family safety net my parents relied on in New York.

For several weeks, my parents scrambled to find an affordable place where we would be safe after school and during the summer. Our new neighborhood's recreation center seemed like their best option. For ten dollars a week, or a little more when there was a trip to the pool or amusement park, my parents knew we had somewhere to be. All they needed to provide was a brown paper lunch with a Capri Sun, sliced fruit, and a ham sandwich. And that brick-and-concrete building, with its large glass windows, sparkling new gymnasium, and rows of foosball tables, became somewhat of a refuge for me.

It was there that I learned how to connect with my peers and nurture what became lifelong friendships, play tennis and ice skate, negotiate conflict, pursue one-sided crushes, and, most important, be a carefree child without any pretenses or shame about the size of my body. We went swimming every Wednesday during the summer, learned calligraphy and pottery, were introduced to new sports, and with the exception of one disastrous trip to a local water park, during which the transportation company that dropped us off forgot to pick us up, we were cared for and safe—just as my parents intended. For five years, there was no place I'd rather be than the rec center, so even now, more than a decade after I was penalized for not following the rules, I still struggle to understand why being kicked out—a small moment in a long childhood— seemed to trigger the mental illness that significantly shaped and altered the remainder of my adolescence.

That day, I grabbed my book bag, bomber jacket, and scarf, and stomped out the glass doors in a rage that threatened to boil over. Tears built in the creases of my eyelids as I quietly walked the two blocks home with my head down and my spirit feeling crushed. When I arrived at our empty second-floor apartment, I pulled off all my clothes, slipped on pajamas, and burrowed under my bedcovers. I curled into a tight ball and cried for hours—through hiccups, a raw throat, and my mother coming home from work, finding me, and rubbing my back and shoulders. I felt cast out of the place where I had been safest. It was the center of my world, but now I believed, as much as I believed that ice cream was a food group, that I would no longer be welcomed. I vowed never to return to the rec center, as my mother tucked me under her arm until I fell into a restless sleep.

My irrational response to what was an ordinary childhood experience should've been a red flag to my parents, but the way I dug my heels into the notion that I would never return to a place where I wasn't wanted didn't initially alarm them. The next day came, and then the day after that, and then the day after that, and I refused to walk into the recreation center and resume my preteen antics with my closest friends. Instead, I'd get off the bus after school and feel a sense of dread come over me as I walked right past the building, and then walked quickly to our apartment, where I'd do homework, watch TV, or engage in anything that would distract me from all the fun I was missing out on. This was a time before smartphones were glued to our palms, so it was impossible to know what my friends were saying or doing without me. Instead, I filled in the blanks with catastrophic scenarios that I couldn't rationalize, not even to myself: "Maybe they'll think I got hit by a car and died," I'd think. "Or maybe they'll think my family moved to a new neighborhood, so I've disappeared forever."

During that first week, my parents pretended not to notice that I was avoiding the rec center, but by the second week, they'd begun bringing it up before I left for school in the morning and during our family dinners. "Are you going back today?" my mother would ask before I got out of the car in the morning. I would shrug noncommittally, though I'd already decided I wasn't returning. Days turned into weeks, weeks turned into months, and the longer I stayed away from the rec center, the more resistant and nervous I became. I started ignoring phone calls and AOL instant messages from my friends because I immediately assumed they'd ask me when I was coming back, and that was a question I just wasn't ready to answer. By the time winter turned to spring, my parents had stopped asking me about returning, and I presumed the matter was settled. If they were okay with me spending the afternoons at home alone, riffling through our small DVD collection and watching the same five movies over and over again, then I'd never need to be in a place where I wasn't wanted.

Then, on an ordinary day in April, my mother picked me up from school, and, without warning, drove me directly to the center. It had been five months, and as soon as we pulled up to the front door, buckets of tears began falling down my face. "Why are we here?" I asked through the tears. "Why?" My mother didn't respond. Instead, she went inside and returned with some of the friends I'd abandoned when I'd decided I was done with that place and its people. My closest friend, Shameika, opened the car door and bent down to hug me, rubbing my back and trying to coax me to take off my seat belt, get out of the car, and walk into the building. I cried and cried and cried until it felt as though I couldn't breathe; that's when I got out of the car. I was choking, crying, and trembling with fear, a fear of facing something that I couldn't understand or even attempt to make sense of. Though at the time, I didn't have the right language to describe the panic attack I was experiencing, I knew it was unlike any feeling I'd had before. I was petrified, and I had no idea why. It took nearly an hour of my friends and my mom reassuring me and reminding me that I was safe—there was no harm coming my way—before I walked through the glass doors again.

That was my first encounter with agoraphobia, a debilitating anxiety disorder that's defined by the *Diagnostic and*

Statistical Manual of Mental Disorders as a "disproportionate fear of public places" that often causes agoraphobics to perceive everything from school to a grocery store as "too open, crowded, or dangerous." Agoraphobia strikes without warning. The disorder is indiscriminate and doesn't appear to have a genetic link. Psychologists don't know why it develops in certain people during puberty and why it strikes others in their twenties, and, even sometimes in their sixties and seventies. When agoraphobia came for me on that day when the counselor asked me to leave, it struck hard and without regard for the normal rhythm of my life until that point. Paranoia, agoraphobia, depression, and anxiety are all close cousins that feed off one another, and they were having a huge feast as I'd anguished about the day when I'd have to leave the house.

After I'd returned to the rec center and began attending every day after school again, my parents breathed a sigh of relief. Whatever was happening with me mentally and emotionally was seemingly over, and I'd returned to being the active, curious, and happy child they'd raised. But agoraphobia, which happens in episodes and is typically triggered by something specific, was still lurking. Though we didn't yet know what to call the illness or how it manifested, I had all the symptoms: intense isolation coupled with severe depression, unexplainable panic attacks, fear of the questions that come with absence, suicidal ideation, boiling emotion—all of it. It would take another two years to be diagnosed, and, once again, it would strike as hard as it could.

Agoraphobia smothered me again in the seventh grade. Every Saturday for three months, my mother and I would trek to a high school auditorium so I could rehearse with the all-city choir, a group I was selected for after a grueling series of auditions, but the actual competition was scheduled to take place during the school day. I was prepared: I'd avoided acidic drinks and foods to preserve my voice, turned in all my pending assignments to ensure I'd have the GPA to be eligible, and packed my choir robe and other garments in a small travel bag. Then, just as I began putting away my textbook to leave my class, as I'd told my teacher I would, he stopped me. "You're not leaving my class early," he said in a hostile tone that I've never forgotten. "You haven't been paying enough attention to today's lesson." When I attempted to stand up and leave, he put his hand on my shoulder and sat me back down in the chair.

I could feel my classmates' eyes on me—on our exchange—as my teacher attempted to wield his power against a seventh grader. I sat back down and began to silently cry, tears streaking my cheeks, as I thought about all my hard work going down the drain. When the bell rang, I grabbed my bag and ran toward the school's entrance, but the bus had already departed with my choirmates in tow. There would be no all-city choir for me.

Preteens are notorious for slipping into temporary funks when they're disappointed; there's nothing abnormal about that, but my outsized reaction didn't at all match the perceived offense. My science teacher's refusal to let me leave class triggered something far worse than I imagine he intended. I no longer wanted to be in school, especially among teachers who would so willingly strip away something so important to me.

I spent the entire afternoon crying in the front office, and when my parents picked me up, I told them, "You can't make me go back to school *ever* again."

When agoraphobia is portrayed on-screen, in films like *Where'd You Go, Bernadette* and *The Woman in the Window*, it's often depicted as an eccentric ailment characterized by un-earned paranoia and a retreat from an esteemed life. Though that's somewhat accurate—and what's more interesting than a world-famous recluse, like the character Bernadette, re-discovering her passion for life after being locked away for more than a decade—these portrayals never consider the loss of control agoraphobics have over our mental state. Teenage agoraphobics in particular are reliant on advocates, including their parents, to ensure they receive the care they need. Much like the fatphobia that dictated so much of how I understood myself and my body, mental illness taught me that there were aspects of my existence that I just couldn't control. I couldn't dictate how my peers, my teachers, and my doctors interacted with my fat body, and I couldn't determine if I'd received the kind of mental-health care that would help me better cope with my condition.

My first therapist, whom my parents found after I began having panic attacks nearly every morning before school, made returning to school full-time and taking medication con-ditions of continuing care, though she seemingly understood that I was at the whims of an illness that I had little control over. After I spent two weeks on a medication that alleviated the panic attacks but turned me into a zombie incapable of functioning in any capacity, my parents decided they'd rather have an agoraphobic child than a child with a blank stare. In an odd twist of fate, my parents opted to change therapists rather than adhere to demands that I either couldn't or wouldn't be able to follow.

THEIR ADVOCACY FOR ME AND MY MENTAL HEALTH BESTOWED ME WITH an agency that carried over into how I navigated other elements of agoraphobia. Each time I met with one of my high school's guidance counselors, I pushed to be transferred to an alternative school where my agoraphobia would be accommodated rather than penalized. But there were consequences for my family's refusal to comply with the therapist's medication recommendation. The school's administrators sent me to truancy court—attempting to terrify my family into forcing me to go to school, panic attacks and depression be damned. Every six weeks, my father would request time off from work, pull a button-down shirt and pair of slacks out of his closet, and accompany me to court. He'd always have an armful of documents in a manila folder, tools to aid him in his quest to prevent me, him, and my mom from being incarcerated for truancy. Each time we escaped that courtroom without being in handcuffs, he'd hug me, tell me he was proud of me for leaving the house for the hearing, and take me to the courtroom cafeteria to have a tuna melt, fries, and an ice-cold Sprite. We'd breathe a sigh of relief knowing that this time we were successful.

His insistence paid off: When I asked the vice principal to compile weekly homework packets that I could complete at home and return to teachers through email, he eventually relented and agreed to the accommodation. When the anxiety medication made me even more sluggish than usual, and varying doses didn't seem to ease the problem, I asked my parents and my counselors to take me off all meds permanently—and they agreed. Being able to advocate for my needs allowed me to exert some sense of control over the worst time in my life. Though agoraphobia still struck without warning or reason, knowing that I had a say in how to treat it helped me imagine a future free of the illness, especially when I was holed up in my tiny bedroom, where the blinds were always drawn shut.

I had my final agoraphobic episode at the age of seventeen, after dropping out of high school, getting a GED, and deciding I was moving across the country to go to college, but my ongoing battle for correct diagnoses and proper care was just beginning. When I was a child, no bigger than the length of my mother's forearm, I developed an ear infection while we were visiting my father's family in Detroit. The story of my ear infection has become a family legend, told and retold at family dinners, barbecues, and reunions. Instead of crying from the pain, I cooed and smiled at my parents—almost as if I were calming them as their concern about the infection clogging my ear cavities grew.

Eventually, at the behest of my grandfather, my parents drove me to the emergency room, where a doctor promptly diagnosed me with an ear infection, prescribed the proper medication to cure it, and sent the smiling, cooing, completely-unbothered-by-the-pain baby home. My parents tell this story often, always smiling at the same moment. "She didn't even cry," they exclaim. "She was the happiest baby we'd ever seen." I also, apparently, didn't cry when a doctor pierced my ears when I was nine months old. I didn't—and still don't—cry when doctors poke and prod, and insist on blood tests, though nurses often insist on drawing blood from the stubborn veins in my arms instead of from my hand as I request.

For many years, I embodied that joyfulness in doctors' offices because I didn't associate medical care with emotional, mental, or physical pain. I cracked jokes with a nurse after I almost sliced my pinkie off in a meat slicer while working at a deli counter. I made the anesthesiologist smile as I was being prepped for a surgery to remove a baseball-sized fibroid from my uterus. When I was in the hospital for an asthma attack that nearly killed me, a nurse brought me a Popsicle because she found my energy "radiating." I figured that humor and happiness could be the antidotes to the anxiety that often seems to hover around emergency rooms and hospitals. As death lingers in the halls, just waiting to claim its next person, and families worry about how they'll cover exorbitant medical costs, pretending that I'm oblivious to my own mortality has always seemed like a good idea.

For a long time, this approach worked and gave me equal power in an unequal power exchange with doctors—until those very same doctors began treating me as a less-than-ideal patient who was making their jobs more difficult merely because of the size of my body. Though doctors have been obsessed with my weight since I was nine, I mostly overlooked their bias until I was in college and began to attend appointments on my own. Every time I visit the doctor's office, it goes something like this: I describe my symptoms; the doctor asks me about my family's medical history; the doctor examines my body, paying particular attention to wherever I'm feeling pain; and finally, the doctor prescribes me some sort of medication before lowering her voice to lecture me about the importance of losing weight. The next step is always referring me to a nutritionist, setting up blood work to test for cholesterol and diabetes, and then calling to let me know that I'm one of the "healthiest patients" they've ever seen.

In my early adulthood, I'd decided to visit the doctor after weeks of suffering from swollen ankles, unrelenting lower-back pain, hot flashes, and uncontrollable weight gain after losing what some people would consider a significant amount of weight. I'd just relocated from Denver to New York City for a dream job in digital media, and between happy hours, after-hours media events, working, and commuting, it felt as if I were running a marathon with no finish line in sight.

At first, I thought the aches and pains would subside once my body adjusted to the hectic pace of a large city, but the symptoms persisted for weeks and then months. I chose this particular doctor because she was a Black woman with the same last name as me; in hindsight, that was a shallow decision-making process, but any sense of connection has always eased my discomfort whenever I visit a new doctor. Choosing her seemed right. Our appointment started off nicely enough. She listened patiently as I rattled off my ever-growing list of medical concerns. Then, she weighed me, asked me for my family's medical history, and promptly diagnosed me as obese.

She never considered my symptoms outside of the number

on the scale. Instead, she reduced all my aches, pains, and hot flashes to the weight I'd gained since moving to New York. "You need to lose weight," she said condescendingly. "Your ankles are swollen from edema," she added as she felt my ankle, squeezing it in places where fluid pooled. "If you don't lose weight, you're going to have early osteoporosis. Your bones will break down."

She then recommended a list of nutritionists who could help me "temporarily eliminate carbs from my diet" so that I could quickly shed the pounds that were contributing to my poor, deteriorating health. She also encouraged me to hire a personal trainer and said that she'd monitor my 100-pound weight loss over the course of nine months. Throughout the tense exchange, however, she never considered my symptoms as separate from my weight, gauged my weight as a symptom of a larger medical issue, or devoted time to figuring out my medical ailment. I decided to get a second opinion. My new doctor attributed my symptoms to several medical issues, including extremely low iron and a vitamin D deficiency. None of my medical issues were caused by what the first doctor diagnosed as obesity. In fact, the second doctor told me that weight gain was a side effect of a larger problem, not the problem itself.

Three years later, I was being diagnosed with heart failure—and learned I'd been exhibiting the symptoms of the disease that entire time. Extreme fatigue, swelling ankles, aches and pains, and shortness of breath are all signs of a heart-related ailment; had that doctor in 2016 sent me for diagnostic tests, she might have learned that heart failure was the underlying cause of the weight gain she was so concerned with. Unfortunately, my experience isn't an anomaly. Many other fat people have experienced varying levels of neglect as well.

On May 11, 2018, Ellen Maud Bennett, described in her obituary as a daughter, sister, and aunt, died from cancer in Victoria, British Columbia. She'd learned that her cancer was terminal mere days before she exhaled for the final time, but according to her family, she'd known she was ill for several years. Each time she sought professional medical treatment, however, doctors dismissed her concerns and told her that losing weight would eliminate her symptoms. "A final message Ellen wanted to share was about the fat shaming she endured from the medical profession. Over the past few years of feeling unwell she sought out medical intervention and no one offered any support or suggestions beyond weight loss," her obituary read. "Ellen's dying wish was that women of size make her death matter by advocating strongly for their health and not accepting that fat is the only relevant health issue."

While it's impossible to know if Bennett could have been saved, her death is further evidence that fatphobia in the medical profession is preventing fat people from receiving quality care. In June 2018, writer Roxane Gay tweeted about doctors not listening to her concerns about her health. "I have tried health providers in five states over 20 years and have not found one that listens," she posted. "And it is incredibly aggravating that people think this is just a matter of finding the right doctor when most doctors are incredibly fatphobic." In 2016, dozens of people of size used the hashtag #fatsidestories on Twitter to share the various ways in which they had experienced fatphobia, including from their doctors. One Twitter user tweeted, "When your doctor says, 'You're much too pretty of a girl to be carrying around all that weight.'" Another shared that she went to a doctor for a sinus infection and was told, "Maybe don't eat what looks good and have a salad instead."

Bennett's tragic story, and the painful stories that these women shared on social media, reveal the depth of what Sasha Ottey, founder of the Atlanta-based nonprofit PCOS Challenge, calls "health-care gaslighting." While Ottey coined the term in relation to how doctors dismiss women's pain in particular—to dangerous and sometimes deadly ends—fat people are also subjected to health-care gaslighting each time doctors assume that our symptoms can be alleviated by losing weight and therefore decide to withhold a proper diagnosis and treatment.

Rebecca Hiles began having intense coughing fits when she was seventeen; eventually, the coughing became so bad that she lost control of her bladder, was forced to wear adult diapers, and experienced constant vomiting. For an April 2018 story in *Cosmopolitan*, Hiles told reporter Maya Dusenbery that she spent many nights in the shower, hoping that the steam would ease whatever was irritating her lungs.

Despite her worsening condition, doctors continually told her that she'd have fewer coughing fits if she lost weight. It took five years and switching to a new primary-care physician for Hiles to be referred to a pulmonologist, who found a tumor in her bronchial tube. Hiles later had surgery to remove her entire left lung, which might have been saved, had she received a proper diagnosis earlier. "Were it not for physicians who tried to treat something beyond my fat, physicians who saw my whole health rather than make assumptions in an otherwise healthy 20-something, I would still have cancer inside of me, I would still be sick," she wrote on her blog, the Frisky Fairy, in June 2015. "I would still be sitting in showers at night, coughing and vomiting. Or, the cancer would have spread and metastasized inside of me. Were it not for physicians that saw me as a person, rather than seeing my fat as something to treat, I very well may have stopped seeking medical treatment, thinking that I was simply mentally ill. I could have died."

Unfortunately, Bennett's and Hiles's cases are more common than they should be: many fat people, women, trans people, and nonbinary people are prejudged by medical professionals, which often leads to mistreatment and neglect. In 1982, physicians were mailed anonymous questionnaires that asked them to select five issues that they negatively associate with a patient's overall understanding of health. One-third of respondents chose obesity as one of those issues, making it fourth behind drug addiction, alcoholism, and mental illness. Furthermore, those physicians associated obesity, specifically, with poor hygiene, noncompliance, hostility, and dishonesty. A 1987 study that surveyed 318 family physicians concluded that two-thirds of those doctors believed their obese patients lacked self-control, and 39 percent believed that their patients were lazy.

Time hasn't improved doctors' attitudes toward fat patients. A 2003 survey of 620 primary-care physicians found that more than half of them viewed obese patients as "awkward, unattractive, ugly, and noncompliant," and that the respondents considered physical inactivity to be the biggest cause of obesity. It's impossible to fully quantify the impact that these biases have on the long-term health of people of size. What we do know is that fat patients are more reluctant to seek medical treatment because they fear that their doctor will shame them.

"Disrespectful treatment and medical fat shaming, in an attempt to motivate people to change their behavior, is stressful and can cause patients to delay health care seeking or avoid interacting with providers," Joan Chrisler, a psychology professor at Connecticut College, said during a panel at the annual convention of the American Psychological Association in 2017. "Recommending different treatments for patients with the same condition based on their weight is unethical and a form of malpractice," Chrisler continued. "Research has shown that doctors repeatedly advise weight loss for fat patients while recommending CAT scans, blood work or physical therapy for other, average weight patients." And a study published in 2015 in *Obesity Reviews* found that there's a significant correlation between how a fat person's been treated by doctors and their willingness to seek treatment.

Some patients avoid seeking clinical care; withdraw from the encounter, making them incapable of advocating for themselves; and, most important, don't get the recommended screenings that could prevent diseases from advancing. For instance, a 2006 study published in the *International Journal of Obesity* looked at the barriers preventing white and Black plus-size women from seeking adequate gynecological care. More than 90 percent of the women observed had health insurance that covered preventative care. The researchers examined women in different body mass index (BMI) categories to determine how their weight influenced how readily they sought medical treatment. The results were alarming: 68 percent of women with a BMI over 55, which makes them technically obese, delayed their Pap smears out of fear of being fat-shamed. "These barriers [included] disrespectful treatment, embarrassment at being weighed, negative attitudes of providers, unsolicited advice to lose weight, and medical equipment that was too small to be functional," the study's researchers concluded. "Obese women reported that they delay cancer-screening tests and perceive that their weight is a barrier to obtaining appropriate health care."

While obesity is often considered the cause of illnesses like heart disease and diabetes, sociologist Natalie Boero argues that this has more to do with the way doctors approach treating fat patients than how diseases manifest in fat bodies: "Who are doctors surveilling? Who are they testing?" In 2009, Rebecca M. Puhl and Chelsea A. Heuer, who are both scholars at the Rudd Center for Food Policy & Health, surveyed 2,449 fat men and women for a study on fat bias among health-care professionals. They found that 69 percent of those surveyed had experienced bias from doctors while 52 percent endured recurring fat bias. As Puhl later told *The New York Times*, fat-shaming, especially by doctors, is "very harmful to health." Boero told me that a focus on obesity is causing doctors to fail fat patients. "We're so caught up in conflating weight and health, but there's plenty of evidence to show that weight and health are not correlated," she said. "Every illness that a fat person gets, you also see it in thin people." By treating fatness instead of diagnosing illnesses and assigning treatments, physicians are putting fat people into impossible situations: Do we lose weight and hope that will solve our ailment? Or do we suffer in silence, afraid of being mistreated by medical professionals?

After years of cycling through this humiliating routine, I'd come to expect—and even prepare—to be mistreated by primary-care physicians, but I wasn't prepared to battle with gynecologists. On a brittle and windy Chicago day in November 2017, my period lasted longer than usual. Day seven became day eight, and the bleeding showed no signs of stopping. I bled and bled and bled and bled for twenty-eight days before I finally decided to visit a gynecologist. I walked into the sterile, white room, already knowing that a fibroid—a type of benign tumor that can grow within and outside of the uterus—was causing the problem. I'd had one fibroid removed in 2015 through a robotic myomectomy, an outpatient surgery that removes the tumor without damaging the uterus. I expected my new gynecologist to agree with my assessment, schedule an intravaginal ultrasound to measure the fibroid, and approve a second myomectomy. Instead, she referred me to my insurance company's metabolic weight-loss clinic to consider either taking appetite suppressants or having weight-loss surgery.

Rather than consider me a steward of my own health, competent and capable of making sound decisions about the best course of treatment, my gynecologist—like other doctors before her—dismissed the pain that brought me to tears and left me nearly bedridden. Bleeding through a menstrual pad every two hours had side effects that no doctor addressed, such as iron levels so low that I couldn't make it down the street without losing my breath. I had no doubt that removing the fibroid would eliminate these symptoms, but after multiple intravaginal ultrasounds and uterine biopsies, my gynecologist prescribed megestrol acetate, a pill that injects the hormone progesterone directly into my body. It was supposed to temporarily stop the excess bleeding and help my body regulate its menstrual cycle. I used megestrol acetate for two three-month intervals, and each time, it seemed to work; I was no longer bleeding through my pants—or bleeding at all—and it allowed me to return to some semblance of a normal life. It was also a temporary solution that controlled the symptoms but didn't solve the problem.

By January 2019, I'd finished my second round of the drug and had been bleeding for more than a month, and the pain and exhaustion were keeping me in bed for many days and many nights. Nothing that my gynecologist proposed worked; I kept getting sicker and sicker. It had taken more than a year, but I'd finally convinced her that I needed to have the fibroid removed. But there was a final step: getting approval from a surgeon. I always need a witness during encounters with doctors, so I brought my mom along. We were both nervous but ready to face whatever came next. When my soon-to-be surgeon walked into the room with her white coat on, she turned her steely blue eyes directly on me. "So, you want to get your fibroid removed, huh? What seems to be the problem?" After I'd recounted all of my problems, she—like my gynecologist before—attempted to blame my symptoms on my size. "Have you ever considered losing weight?" she asked me, immediately causing me to look downward at the floor. "If not, maybe you should lose fifty pounds and see if that will help with the bleeding. We can also install an IUD." I felt as small or smaller than I'd ever felt before.

Thankfully, my ever-vocal mother, who would later have a hysterectomy to remove fibroids, stepped in. "She needs to have the surgery—immediately," my mother retorted. "My daughter is tired. It's enough already! She's done everything her doctor has asked, and it isn't working." The physician relented and agreed to perform the surgery, but as a fat Black woman, I had already heard her message loud and clear. We know little and they know it all. It's a familiar routine that I've been encountering for more than twenty years—and it never seems to get better. When the surgeon went in to remove the fibroid, she discovered that there were actually five—four of which were hiding behind the larger fibroid—and there had been little chance of medication or an IUD ever curbing the symptoms. We needed to excise the problem from its root. Though there are a number of abstract reasons that doctors dismiss fat patients—including deep-seated fatphobia, and, for those with multiple marginalized identities, racism, transphobia, and classism—there's also a more logical reason: fear is a marketing ploy.

In many ways, fear is the point. Doctors, dieticians, and companies that peddle weight-loss supplements and expensive exercise equipment are invested in both promoting and upholding an unattainable thin ideal—and they're willing to stoop so low as to scare fat people into buying whatever it is they're selling. The medical establishment's narrow-minded focus on thinness is the only reason that physicians still rely on BMI to determine a person's overall health. But BMI has long been proven to be inaccurate and misleading.

In an August 2013 study published in *Science*, University of Pennsylvania researchers Mitchell Lazar and Rexford Ahima found that people with a BMI over 30 can have a better overall metabolic profile and lower mortality risk than those who fall within the normal BMI range of 18.5 to 24.9. Lazar and Ahima have called for the BMI to be modified to include "accurate, practical, and affordable tools" that "measure fat and skeletal muscle, and biomarkers that can better predict the risks of diseases and mortality." Doctors treat BMI as if it's the holy grail, but without considering factors such as age, fitness, preexisting diseases, and genetics, it's a flawed measure of health at best.

The BMI also doesn't consider how fat is distributed through the body, according to Margaret Ashwell, an independent consultant who used to be the science director of the British Nutrition Foundation. In 2012, Ashwell noted that measuring waist circumference is a better indicator of future chronic illness than BMI, because BMI doesn't consider where fat is concentrated, which is a significant factor when assessing risk for certain diseases. And, of course, BMI takes into account only overall body weight, making no distinction between muscle and body fat. When I was diagnosed with heart failure, my cardiologist mentioned several factors that could've caused the disease; obesity wasn't one of them. If doctors scare patients into seeing one of their nutritionists, paying an exorbitant amount of money for weight-loss supplements and group counseling sessions, and hiring an insurance-approved trainer, everybody wins—and by everybody, I mean doctors and insurance companies.

None of this surprises Natalie Boero, the sociologist whom I spoke to in 2016. She and her daughter have been repeatedly fat-shamed by doctors. When a children's hospital admitted her seven-month-old daughter for a respiratory infection, they weighed the baby and declared that she needed nutritional counseling. "[A nutritional counselor] looks at me, she looks at my daughter who, at that point, is a roly-poly baby. And she says that my daughter's file has been flagged for nutritional counseling because of her weight," Boero said. "She starts asking how much juice I'm giving her. I'm not giving her juice at seven months old."

Boero asked the nutritional counselor to leave, but she realized later how common this experience is. "This is the first contact parents get with this: their baby is being [fat-shamed]," she said. "She came through the ER with a respiratory infection." None of this actually improves the health of fat people, which is why it's ineffective. "If you cared about my health as a fat person, you wouldn't fat-shame me," Boero said. "You wouldn't assume I'm in bad health, and you would actually look at what things are doable that can have lasting impact. You would make sure I could get care."

So, how can fat patients ensure that their doctors are treating their ailments instead of their size? Boero says that it's essential for fat patients to be proactive in seeking second opinions. Don't settle for the opinion of one doctor; keep seeing other physicians until you get a diagnosis and treatment plan that alleviates your symptoms. Patients should also seek out doctors who subscribe to Health at Every Size, which encourages health-care professionals not to make diagnoses based on weight or to encourage patients to lose weight. Patients can also request not to be weighed, decline to have a conversation about weight loss, and ask questions until they feel satisfied with the diagnosis and the treatment offered. And, most important, no one should be afraid to tell a doctor when they're not being helpful or are perpetuating weight stigma.

Boero also suggests using the National Association to Advance Fat Acceptance's online resource about how to communicate effectively with doctors and recommends setting boundaries with doctors before going in for appointments. "Being educated about fat-shaming, sharing experiences on this, and emailing doctors [about Health at Every Size] are great resources," she said. "I really think the best thing to do is contact the doctor before. Tell them that you want behavioral changes that can help, but you want them to treat your problems, not your size. [Tell them,] 'I'm not really open to weight loss or prescribed weight loss.'"

Advocating for yourself with your doctor really can get

you the treatment that you deserve. My doctor's fat-shaming approach depressed me. It made me feel inadequate, small, and unworthy of medical care. Now, I'm more proactive about my care, and I use all the tools at my disposal to advocate for myself. But I shouldn't have to. I hope that our doctors might one day learn to care about our health—not our weight. Our lives literally depend on it.

IN THE LAND OF PRINCESSES, ALL WE DO IS CATFISH

When I was a child, my parents split their schedules so that at least one of them would be home with my brother and me. Since my father worked at night, he was tasked with entertaining us in the afternoons we weren't at the rec center. So, one summer, he took us fishing at one of our neighborhood lakes nearly every day. We went so often that the routine was locked into our minds. My brother and I would get ourselves dressed; either help my father pack us lunches—ham sandwiches that became sweet over time in our lunch pails, Capri Suns in pouches or metallic cans, and at least two pieces of fruit—or talk him into buying us lunch; grab the white Styrofoam container of worms from the refrigerator; and carry all of our supplies, including our plastic tackle box full of bobs and lures, to the car. Every day, we'd embark on an adventure, thrumming with the thrill of both something new and something constant.

Sometimes, our father's friends would accompany us. Other times, we'd drive a couple of hours to a far-off reservoir where we'd be the only people casting our poles into the dark brown—or, if we were lucky, bluish-green—water. We arrived with a single mission in mind: catching a fish, admiring it, and then tossing it back. There's something calming about the practice of fishing, though there are no guarantees that a trip will bear fruit. Oftentimes, we'd leave the lake—hot, tired, and completely spent—without ever getting a single fish on our hooks. But fishing is a discipline; actually catching fish is beside the point for those who aren't professional fishermen. It taught me and my brother patience. Fishing taught us to try again, employing different methods after our initial attempt failed. It taught us to be quiet, to be okay with stillness, and to be okay with the art itself being enough to sustain us.

Within all this, however, there was nothing more thrilling than seeing my bobber drop beneath the surface, signaling that a fish had found its way to my hook. Reeling the fish in, attempting to avoid jagged rocks and prevent the fish from breaking the line, is a singular joy, and when the fish cracked the surface—meaning it was hooked and couldn't escape—a smile would light up my face. I'd done it. I'd achieved the goal.

When my own father was a kid, his family in Detroit had spent their days off piling into cars—buckets of fried chicken and containers of potato salad in coolers between their feet— and driving to picnic and fish on Belle Isle and Kent Lake. We inherited that family tradition, carrying it from his generation to ours, and by the end of that summer, the lake was one of the places my brother and I loved most. Though I don't fish nearly as much as I used to, fishing is in our family's blood.

THE LIBRARY WAS MY FIRST FORMAL INTRODUCTION TO THE INTERNET. THIS might seem confusing to those born after 1993, but there was a time in which most households, particularly those as middle-class as mine, didn't have internet access. In-home internet was both a rarity and a luxury. After my family left Queens for Denver, there was a community center nearby that included our beloved rec center, a fire station, soccer and tennis fields, and, of course, a small library. Our neighborhood library was at the top of an eighteen-step stoop and down a long, dark hallway lined with two water fountains and two bathrooms. The shelves were dark brown and dusty, but they were ours, and nearly every afternoon after school and every weekend, my friends and I would leave the rec center and walk across the square to the library. We had to create our own fun, and we spent hours browsing the shelves, picking up books, putting them down, reading a few pages, and chatting with the librarians, who became like surrogate mothers for many of us. They recommended books, kept an eye on us, and even watched us as we walked home.

There were some things those librarians couldn't see, though. One day, we walked in to find that a bay of six computers, separated by clear dividers, had magically appeared in the front of the library. There was another bay near the center, and a third in the back near the bathrooms. These computers had access to a thing called the "internet." My friends and I looked at one another in wonderment when we realized that we could sign up for hour-long slots. At first, it was taboo to search the internet alone, so one of my friends would sign up, and then three or four of us would huddle around the stall, pulling up chairs and peeking over one another's shoulders as we clicked through AOL's home page, venturing onto various new websites that we learned about by listening to the browsing adults around us.

As teenagers, it had always felt that our world was confined to just a few square miles. But when one of my friends introduced us to chat rooms, it changed the whole game for us. He described them as secret places where we could talk to one another, no matter where we were in the library, and we could also talk to strangers who maybe lived in the same city and maybe lived somewhere else. It seemed as if he was revealing some mythic, impossible folklore. I can recall looking at him, my jaw slack, as he excitedly showed us how to create silly user names, log into these chat rooms, and then begin talking to other people. At first, I didn't understand the premise, but once I started using chat rooms—first only talking among people I already knew and then broadening my scope to strangers—I felt an almost surreal surge of adrenaline. I began joining specific rooms that aligned with my interests: reading The Baby-Sitters Club and Sweet Valley High, watching WCW and WWF, and obsessing over Disney princesses. I could meet new people from all over the world and build connections with them. But, most important, I could be whomever I wanted to be.

At first, I didn't realize the importance of using aliases

online. This was before Chris Hansen dominated prime-time news with his *To Catch a Predator* specials, and before parents across the United States became deeply concerned about the influential power of the internet over their children. It was 1998. I was nine. I didn't see the inherent danger in being myself online; it seemed like the right thing to do—a means of projecting authenticity, even if it came at my own expense. At that time, chat rooms required users to offer a first and a last name, an address, and an age. Though I fudged the age—you had to be seventeen to join, after all—I used my real name and my real address. While there were many Whitney Houstons, Sean Combses, and Mariah Careys floating around, I was often the only Evette, and I would detail actual real-life events to strangers.

One time, I even allowed a stranger to call me on my home phone. I waited until my parents and my brother left our apartment to grab lunch, and then I chatted that he could ring me. We talked for ten minutes, maybe less, and in hindsight, it was all incredibly gross: his voice was as deep as my father's, so I knew he was a grown man, and he asked me about what I was wearing, if I was home alone, and what a young girl like myself was doing in the chat room. It seemed innocent enough—just a chat conversation brought to life—but our intentions weren't at all aligned. I wanted to discuss the latest episodes of *Rugrats* and *The Wild Thornberrys*; he wanted to talk sexily to an eleven-year-old on the phone.

I know, I know. I could've landed on an episode of one of those eerie shows that Investigation Discovery airs all day, every day, but this was a different time. Honesty isn't currency on the internet, though it is presumed to be. We all project avatars—a representative of our very best self that doesn't encompass the fullness of who we are. We're giving strangers the parts of our lives that feel worthy of shine, while simultaneously scuttling our lowest points into the darkest parts of our closets, pulling them out only when our online persona is seemingly struggling to be its most engaged self. Hindsight helped me to learn these lessons—but so did *Catfish*, the MTV docuseries I've been obsessed with since its inception.

Catfish, in its way, represents the very best and the very worst tendencies of the internet—from deception to uncovering said deception to building community in spite of the alienation that comes from being lied to. The show, which originated as a movie about host Nev Schulman's own experience being catfished, has a simple premise: Someone is in some kind of entanglement with someone they've never met. Sometimes, it's a financial relationship—as is the case in one of my favorite episodes, about a fraudulent music producer scamming strangers out of thousands of dollars. Other times, it's a familial relationship or romantic relationship. In all these instances, there's a person who has met someone online, never interacted with them in person, and encountered numerous obstacles as they attempt to make a face-to-face meeting happen. Hilarity ensues.

One of the show's most memorable episodes focuses on two cousins. Carmen calls Nev and his cohost, Max Joseph, because she's worried that her cousin, Antwane, is being catfished by his alleged boyfriend. All the telltale catfishing signs are there: Antwane has never videochatted with Tony or met him in person; whenever Antwane suggests that they meet, Tony finds an implausible excuse for being unable to do so; and Tony has sent still photos only, never videos or anything that shows what he's doing in real time. In typical *Catfish* fashion, Nev and Max unravel the mystery, discovering early on that Tony is using pictures that he stole from someone online. But they can't figure out who exactly Tony is. They attempt to track him down, driving from house to house—until Carmen drops a bombshell on a wintry afternoon in Cincinnati as the whole crew stands together looking around. "I'm Tony," she says, as her cousin looks bewildered and stunned. "I'm Tony. You should've never called me a fat-ass Kelly Price!" It's by far the greatest moment in *Catfish* history, and it's also the most instructive. Carmen catfished her cousin for two years— deceiving him, talking him into falling in love, and keeping him on the end of her line—because he'd insulted the size of her body. Although Nev and Max attempt to repair their relationship, trying to understand what caused the breakdown, it's clear that something has been irrevocably tripped—and it started with a fat joke.

Catfish has been airing for a decade now. The show's scammers have struggled with a number of issues, from gender identity to homelessness. But I'd estimate, unscientifically, of course, that 85 percent of those who catfish have been fat. One of them was so fat, in fact, that she catfished two people on the show and then, for good measure, appeared on *My 600 lb. Life*. Go figure. It's a cycle of trauma wrapped in the disguise of humor and ridiculous tactics. But fat people who are exposed as catfish on the show typically have a script that's so common that it seems fake: I was bullied and I felt unloved. I didn't have anywhere to turn. I created a fake account because I wanted to talk to people who could possibly understand and give me comfort, but nobody would talk to me because I'm fat. So I decided to portray myself as someone else, and the same people who bypassed me were suddenly interested in me. I planned to tell the other person I was using fake pictures, but I was worried about being abandoned—again—so I lied until I was forced to tell the truth. Sometimes, this story, which varies but usually has these contours, is accompanied by tears, while other times people who catfish attempt to laugh it off. But underneath their decision to lie is a simple need to be accepted, understood, and supported.

IF YOU'VE SPENT ANY SIZABLE AMOUNT OF TIME WITH CHILDREN IN THE past three years, then you've likely heard them gush excitedly about Roblox. Players create avatars that let them move around a virtual world, playing one-off games and traveling from server to server. These kids beg their parents—and in my case, aunties—to buy them Robux, the currency used on the game to purchase everything from outfit changes to homes. My nieces are obsessed with Roblox, and although they're allowed to communicate only with each other and the classmates they already know in real life, the game itself gives them the illusion that their world is more expansive than it actually is. Roblox is customized to suit the needs of Generation Z, but from the minute I browsed the interface, I was immediately transported to the digital portal of my teenage years: Habbo Hotel.

Habbo is more than twenty years old now, so when I began using it at the age of thirteen, it was an earlier iteration of what it would eventually become—a virtual hub for teenagers across the world to be in relationship with one another, whatever that looked like. Even though Habbo was initially designed with adults in mind, the platform was quickly populated by teens looking to connect with one another. And as you accrued more Habbo bucks, you could upgrade everything about yourself, including your clothing, the games you were able to play, the rooms you were able to venture into, and the internet relationships and circles you were able to join. There was a clear pecking order: female avatars with doe-shaped eyes, either a ponytail or curly hair, and square shoes instead of round shoes were at the very top, followed by male avatars with slouchy hoodies, those same eyes, and square shoes rather than sneakers. Unlike in Roblox, these Habbo features were free, but those who were immersed in the platform knew how to dress their avatars to appear cooler. Seasoned Habbo users would even help our new friends transform their outfits to blend with the "in crowd."

I became Shay, an avatar with caramel-brown skin, a ponytail, and large doe eyes to match. My best friend and I were members of that "it" crew, and we recognized the other cool avatars from their clothes, and across virtual diners and in pools and private rooms. Over time, these avatars became some of my closest internet friends, and we shared things about our lives outside of Habbo. My best friend and I would flit around from room to room, changing outfits and talking to strangers from everywhere from Canada to Puerto Rico.

Our collective dominance was clear: when our avatars walked into a room, other avatars flocked to us, as if we were in a high school cafeteria, and we'd talk about everything and nothing. Virtual popularity gave us the veneer to discuss things we weren't brave enough to discuss in our real lives: mental illness, suicidal ideation, porn. We knew we'd never meet the other avatars in our real lives, so we spent hours having deep conversations about the dark things we really didn't want anybody, particularly our parents, to know about.

When I started planning to run away when I was fourteen, the first person I told was Erick, one of my Habbo boyfriends. He discouraged me at first, telling me he loved me and he wanted me to stay as safe as possible, but as our conversation continued, he realized how serious I was and, in turn, became serious about helping me. We thought through a packing list together, how I'd get money for a bus ticket (from my dad's wallet when he was sleeping), and where I'd go (my cousin's apartment in New York). Though I ditched my plans when I was ten steps away from my apartment door and realized I didn't have enough money for a sandwich, let alone a bus ticket, conspiring with Erick brought us closer together. I could trust him.

If you became "close" enough to your Habbo friends, then you began communicating on AOL Instant Messenger (AIM) and sharing blog posts through LiveJournal. That's where true intimacy happened, including sexting and acting out sensual and aggressive sex. It reminds me now of movies like *Two Can Play That Game*, when Shanté (Vivica A. Fox) is envisioning slapping Conny (Gabrielle Union), but when the camera cuts away, we see she's just standing there as Conny explains why she's at a company party with Shanté's boyfriend. There's what we want to happen, and then there's the reality. If my life were a movie, the cutaway would reveal me sitting on a rickety computer chair, wrapped in a blanket with my hair uncombed, isolated and alone, and glued to the screen, talking to far more people I didn't know than people I knew. At first, Habbo Hotel was an extracurricular activity, one of the few things I had the energy to participate in after a long day at school—and then it became something more valuable than school itself. It became my education. It became an escape route from a life I often wanted to end. It became a salvation. Who cared if I wasn't actually Shay?

THERE CAME A TIME WHEN FEWER AND FEWER MEMBERS OF MY CLIQUE were using Habbo Hotel. We'd migrated from the platform— bored of the mundanity of the experience—and were now communicating through AIM. We swapped our Habbo statuses that detailed who we were dating for AIM away messages, where we offered curated peeks into our mental state: our favorite music lyrics, quotes from books we'd never read, and words that pined for our internet loves. Over time, AIM, like Habbo Hotel, became my lifeline because I could no longer go outside. By that point, I'd developed agoraphobia, so I'd suffer panic attacks when I even attempted to leave the house in the winter. I was as isolated as I'd ever been. I didn't have a cell phone then and neither did my friends, so I couldn't text them about what I was experiencing. But I was probably too ashamed and embarrassed about my struggles to want to talk to them about it anyway. And eventually, thanks to my condition, I dropped out of school to pursue a GED.

I met Dereck on Habbo, and we eventually migrated to AIM. Ordinarily, I would've avoided him because his avatar didn't meet my aesthetic standards, but somehow, we fell into a conversation. Soon we were talking every single day—mostly about the fantasies we'd curated about our lives—until we'd moved from being internet friends to being in a committed relationship, whatever that looks like for two teenagers who lived thousands of miles away from each other. I wonder, now, if Dereck was also agoraphobic. I wonder if he was a he at all. I wonder if Dereck was simply attempting to feel something other than grief, sadness, and a tenuous grip on reality. I wonder if Dereck saw me as I saw him—as a way out. Dereck, it seemed, needed someone in his life to save him from dire circumstances—a life vest who kept him afloat, buoyed by positive affirmations and prayers.

When we first began corresponding, I was thirteen—though I told him I was fifteen—and he said he was seventeen, though I've never been able to confirm his age. He told me about his life, a story so troubled it seemed designed to earn sympathy, although I didn't see it that way at the time. He told me he was living in Atlanta, in an apartment he leased in his mother's name, though his mother wasn't present. Sob. She'd been addicted to drugs, and thanks to the dark places that disease takes people, they'd been evicted so often that he was forced to drop out of high school and begin selling drugs. Sob. If this wasn't sad enough—an urban novel in the making—

Dereck was a teenage father. He and an ex-girlfriend had fraternal twins, Jamal and Jamia, who lived with him because their mother also developed an addiction to crack cocaine. Sob. All of the morally questionable behaviors he described to me—selling drugs, pistol-whipping dealers in his ring who flubbed the profits from the crack he fronted them, using his mother's name on a lease without her consent—seemed excusable and even justifiable; he needed to survive, after all.

By the time we began corresponding, Dereck said he wanted out of the drug game, but there was a steady cast of characters preventing him from starting the new life he'd imagined for himself. Day in and day out, as I sat in my parents' apartment, isolated from the world and exiled by agoraphobia, I followed the intense happenings in Dereck's Atlanta apartment, courtesy of AIM, LiveJournal entries, and MySpace. Web cameras were still relatively rare, and it wasn't unusual to have an internet relationship that never involved even a phone call. You had to trust that the person on the other end of the screen was telling you the truth—that their name, age, birthday, and photos were real and their interest genuine. I bought into it all.

When I created my first MySpace in 2006, a couple of years after Dereck and I started chatting, we made matching pages that declared our love for each other. Our anniversary was September 6, 2006, so he was Clyde 9606, and, you guessed it, I was Bonnie 9606. I was his top friend and he was mine. All day, every day, we would talk—or I would talk to everyone in his orbit. His childhood girlfriend and I would argue about who Dereck loved more. (I was a girl, not yet a woman.) I would bond with his son, Jamal, who was eight at the time, over our shared love of WWE. Dereck would leave the computer, or so he said, and I'd talk to Jamal until it was his bedtime. Then, Dereck and I would resume our conversation. Dereck and I would talk about a future in which I'd move to Atlanta, he would leave the drug game, and we'd move into an apartment close to a college campus, where we would raise his children together as I earned my degree. It was the ultimate fantasy—a means of escaping circumstances in which I felt completely out of control. Our relationship was so "real," in fact, that my mother eventually removed the computer from our home to limit our contact. But that didn't stop me: late at night, I'd sneak my mom's work laptop from her briefcase, guess her password until I cracked it, and instant-message with Dereck in the darkness of my bedroom. As soon as I got back in touch, we fell right back in step. We had a plan, and nobody could interrupt it—except for him.

In hindsight, I realize that Dereck wasn't Dereck. We communicated for more than five years—even after cell phones became more popular and we both owned one—but we never had a single phone conversation. As we planned to meet, his excuses for being unable to come started to escalate—from simply not having transportation to pick me up from the airport to having a stroke. Yes, a stroke. When I learned the news from Jamal via a MySpace message, I was terrified. I loved Dereck. I couldn't imagine losing him, especially in such a sudden fashion. I curled up under my covers and bawled my eyes out, feeling helpless to support the one person I loved the most. Emotion shrouded the logic that should've kicked in: Dereck, a perfectly healthy twenty-one-year-old man with no preexisting conditions, randomly had a stroke mere months before we were set to meet. But I was too wrapped up in him and the world we'd created to recognize that he was deceiving me.

I'd originally planned to attend college in Atlanta to be close to him, but Dereck told me to apply to historically Black colleges and universities in other cities, because that made more sense than us meeting and living together. I can remember the emotional weight of being rejected by someone whom I often imagined lying beside me at night, wrapping his thin, chocolate arm around my waist, and whispering sweet niceties in my ear. It was devastating to be told that the fairy tale we'd been crafting in our private world for more than five years wasn't going to translate into a real-world relationship. Our relationship soured when he still refused to speak on the phone after his alleged stroke; instead, he'd send me misspelled texts, almost as if he were slurring his speech through text message. Couple his refusal to take our relationship offline with the newfound thrills of college, and our relationship was doomed.

Once I began meeting men in real life and pursuing relationships that included some form of physicality, Dereck and I drifted apart. While he knew he no longer had me ensnared on his hook, he would still text me sometimes, almost as if to remind me that we had a past that always lurked in the background of the new life I was building for myself. I was no longer an agoraphobic without a connection to the outside world, but a college student coming out of the mental-illness-induced shell that had allowed him to reel me in time and time again.

We've now been treated to the spectacle of *Catfish* for more than ten years. All the tools the show utilizes—reverse image search, reverse phone number lookup, researching Facebook friends—are accessible to those who aren't professional internet investigators. But there's no shortage of guests who are still being tricked by a stranger—or someone on the periphery of their life who wants their attention—because catfishing is less about the deception and more about the connection. I didn't realize Dereck was a catfish until I watched the show for the first time when I was twenty-three. As guest after guest explained their catfish's increasingly odd excuses for not meeting, I knew in the pit of my gut that I was in the same boat. It was akin to vampires walking in the sun for the first time. At first, it's heavenly to feel the newfound rays on your face, but then that joy turns into dread and then sadness as your body burns from the inside out. I was thrilled to know I hadn't envisioned that gut feeling after his stroke that something was amiss. I was also sad to have my suspicions confirmed.

Suddenly, all his excuses snapped into place and I could see the full picture of his deception. Up until that point, I thought I was the only person who'd experienced this phenomenon; it embarrassed me so much that I told my early college friends that Dereck was real. I invented an entire relationship and shared stories with them that never happened or happened only through the computer screen. My family knew Dereck had been lying to me. They'd been trying to force me to see it for years. My grandma and mom were vocal about their belief that Dereck engaged in sex trafficking of some kind and they'd never see me again if I moved to Atlanta. But once I went to college in Maryland, they never mentioned it again. After all, I had dodged the bullet; there was no reason to continually remind me that I'd almost been shot. When I told my mom that Dereck had catfished me, she simply shook her head. She knew. I just hadn't been able to see it.

I was older, wiser, and had been in multiple real-world relationships, but it still upended me emotionally. I texted him to ask if he'd been catfishing me, but he refused to give me a straight answer. Instead, he told me that he loved me and there was nothing fake about his feelings—a line that catfishes on the show often tell those they've been deceiving. I let go of the fantasy of Dereck at that very moment. Although I was heartbroken, I decided I would never speak to him again. I never did.

When I reflect on the time I spent with Dereck—much of which is a blur, a near blackout that I can remember only in snippets—what strikes me hardest is that I was willing to be catfished because I simply wanted to be in a relationship with someone who could help me escape the realities of my miserable teenage life. I could invent adventures I'd never had—vacations I hadn't been on, friendships I didn't have, afterschool activities and jobs that I'd never pursued because I'd dropped out of high school. My online life was magical, even as my real life fell apart, leaving me to navigate suicidal ideation and heartrending isolation. With the exception of my close family, who genuinely supported me through my condition even if they didn't understand it, I felt that nobody else was there for me. I never lied about my appearance—there was no need to at the time—but I still pulled in people who were attracted to my personality and my perceived online popularity. That's the entire point of catfishing. You're luring someone in—designing an existence that will appeal to them—and then maintaining the attraction by elevating lie after lie after lie.

But what he taught me—just as those childhood fishing trips did—is that sometimes, the connection itself is more important than everything that accompanies it. I just wanted to be loved. I imagine Dereck did, too. Yes, some of us are cold-blooded, but the majority of us want to feel desired, supported, accepted, understood. And even if the person meeting those needs isn't who they pretend to be, isn't it better than being alone? Isn't it?

I WANT A LOVE LIKE KHADIJAH JAMES

*D*uring BET's 2013 Black Girls Rock ceremony, writer, producer, and director Mara Brock Akil—the mastermind behind *Girlfriends*, *The Game*, *Being Mary Jane*, and other shows that explore Black women's varied experiences—explained the difficulty of bringing the kinds of stories she champions to the small screen:

> We're born in a country whose greatest export is images, and that's great, right? Everywhere we turn—museums, TV, movies, magazines, books—we'll see our beauty reflected, unless you're Black or a girl. We walk around in our home called America and we don't see our pictures hanging on the wall. It's like we don't exist. And when there is an image that resembles us, often, upon closer inspection, it's not us.

In the years since Akil's speech, Black women have landed numerous lead roles. Since *Scandal* ended a nearly forty-year drought of Black woman actors leading a prime-time network, we've been treated to *Deception, Sleepy Hollow, How to Get Away with Murder, Black-ish,* and *Empire.* There's also been HBO's *Insecure* and *Lovecraft Country,* Starz's *Run the World,* Amazon Prime's *Harlem,* BET's *Bigger, Twenties,* and *First Wives Club,* and so many others.

While more Black women are springing up on-screen, there's still a glaring dearth of fat Black characters. We've seen fat Black women in significant roles: Natasha Rothwell costarring as Kelli on *Insecure,* Retta (as Ruby Hill) committing crimes every week on NBC's short-lived dramedy *Good Girls,* Queen Latifah playing music manager Carlotta Brown on Fox's canceled drama *Star,* and Nicole Byer costarring as Nicky in NBC's *Grand Crew.* Yet, no fat Black woman is singlehandedly leading a major scripted show.

The absence of fat Black women actors echoes a larger trend that impacts all plus-size actors. In a 1980 study published in the *Journal of Communication,* Lois Kaufman, a former professor in Rutgers University's Department of Humanities and Communications, found that only 12 percent of prime-time TV characters were overweight, and there was an overrepresentation of average-size and undersized TV characters. A 2003 study in the *American Journal of Public Health* revealed that not much had changed: only 14 percent of the 1,018 characters on commercial television between 1999 and 2000 were overweight or "obese," pointing to an "imbalance toward thinner men and, especially, thinner women." The

treatment of fat characters further underscores these findings: when television programs and movies feature heavier
actresses, it's typically in comedic roles where they're teased
about their weight or their weight becomes a punch line.

On July 19, 2018, Netflix released the two-minute trailer
for *Insatiable*, a revenge comedy trafficking in the same formula that turned *Carrie*, *The Craft*, *Heathers*, *The Lost Boys*,
and other high school dramedies into cult classics. Like its
predecessors, *Insatiable* revolves around an ostracized student
who learns to reject feelings of disempowerment and decides
to unleash her rage on her tormentors. In this newest iteration, Patty, played by Debby Ryan in a fat suit, is the shunned
student who endures incessant verbal and physical abuse from
her classmates simply because her body is larger than theirs.
Much like the other high school dramas that have prolifer-
ated in pop culture since the 1980s, *Insatiable* perpetuates the
notion that those who are perceived as different exist at the
bottom of social hierarchies and are thus subjected to bullying by their peers. In *Insatiable*, the ongoing torment reaches
a cruel end when a fellow student punches "fatty Patty" in the
face, causing her jaw to be wired for the duration of summer
vacation.

While having a closed jaw isn't on anybody's summer wish
list, the assault begets an unexpected "blessing": Patty is forced
to go on a liquid diet that causes her to rapidly lose weight. Although dieting can encourage disordered eating behaviors and
usually results in between one-third and two-thirds of people
regaining more weight than they've lost, *Insatiable* doesn't
address these common problems on a show geared toward

teens. Instead, Patty's body is suddenly transformed, and so is her position on the totem pole of high school popularity. But Patty doesn't find solace in her classmates' newfound willingness to befriend her; instead of forgiving their transgressions, she decides to exact revenge on those who've ruined her high school experience. In the two-minute teaser, Patty says in a voice-over, "Having my jaw wired shut lost me more than just my summer vacation. Now, I could be the former fatty who turned into a brain. Or an athlete. Or a princess. No. I'd rather have revenge."

Teen Vogue attempted to frame *Insatiable* in a positive light, making the claim that the show is "a dark revenge comedy taking aim at teenage bullying, society's warped beauty standards, and the monsters that live inside all of us." Ryan added that the show brought attention to the justifiable rage of her character and so many others who have been bullied. People "don't fully understand the depth and breadth and fire of Patty's rage," she told *Teen Vogue*. "And I don't think Patty does, either, so throughout this process she's kind of discovering how far she'll go to either exact revenge on anyone who's wronged her or, in her mind, make things right and give herself all of the things that she was robbed of or feels she was robbed in kind of the earlier first bit of her life."

Sure, revenge is bittersweet for those who've been on the receiving end of physical and verbal torment, but why are fat people always depicted as helpless victims who are either being bullied or bullying others? Think, for a moment, of all the fat characters you've seen on-screen. How many of them are actually able to navigate fatphobia with their boundaries and dignity intact? How many of them are downtrodden and constantly inflicting pain on others?

In 2005, Ryan Reynolds and Amy Smart starred in *Just Friends*, an underrated romantic comedy with a familiar plot line: Chris Brander (played by Reynolds in a fat suit) is in love with his childhood best friend, Jamie Palamino (Smart), but she regards him as a friend instead of a potential boyfriend. When Chris reveals his crush for Jamie during a high school graduation party and one of his bullies humiliates him by reading his sappy love letter to all the guests, he decides to skip their small New England town for good—moving to Los Angeles, becoming a famous record executive, losing more than 100 pounds, and treating women with a sexist disregard. Of course, Chris and Jamie are a loving couple by the end of the movie; after all, it's a romantic comedy. But Chris spends the entire movie eluding his trauma instead of facing it head-on: he desires to exact revenge against those who tormented him during high school, delighting in their emotional and financial misfortunes, and using his newfound weight loss and power in the music industry to humiliate his former foes.

If fat (or formerly fat) characters aren't vengeful manipulators, they're often helpless misfits whose main personality trait is self-hatred of their body. *This Is Us*, one of those classic shows that stays with you long after the season has ended, has a single fat character, Kate Pearson (Chrissy Metz). She feels inadequate in comparison to her successful family members and resentful of her mother (played by Mandy Moore), who feeds Kate cantaloupe for breakfast and encourages her to wear a T-shirt over her swimming suit. We see Kate become obsessive about her weight, in a spiral of yo-yo dieting. Given that fat women have few representations, the mere existence of Kate—who doesn't have an hourglass figure or a size-12 frame—is considered "revolutionary." But she exemplifies all the worst aspects of the "tragic fat girl" character, a repeat of the script we've seen time and time again.

Fat characters are often pigeonholed into several stereotypical roles: the depressed and bullied protagonist who believes that happiness will come only once she's thinner; the sassy sidekick who provides emotional support for a thinner character; the asexual mammy who is too focused on taking care of other people to pursue her own dreams; and the evil bitch who uses irrational cruelty as a protective measure. We've had *Huge*, an ABC family drama about seven teens who are sent to an outdoors weight-loss facility to slim their bodies down and lose some of their emotional and mental baggage as well. We've had Kirstie Alley playing a version of herself on Showtime's *Fat Actress*, as she struggles to lose weight and gain control over her career. They're obsessed with losing weight. They feel slighted or stunted in their careers. They're nurturing of everybody in their orbit, but nobody nurtures them.

In movies such as *Pitch Perfect*, *Rough Night*, *The Heat*, and *Bridesmaids*, plus-size characters aren't afforded the opportunity to fall in love. They're relegated to sidekicks who break up tense moments with exaggerated humor that usually pokes fun at the size of their bodies. This bleeds into how we're engaged with in real life, and enables our partners, including my college boyfriends, to mistreat us without consequence. If fat bodies are meant to be ridiculed, shamed, and hidden, then there must be something wrong with being attracted to a fat person. Hide us away, and the problem's solved.

Insatiable, which takes the "depressed and bullied protagonist" trope to an extreme, can't reinforce every single negative stereotype about fat girls and still consider itself a positive representation of fat teens. As librarian and fat-acceptance activist Angie Manfredi tweeted after the trailer was released, "You have a chance to make creative, engaging, original content from fat people about fat people's lives and you choose to put a skinny person in a fat suit and make jokes about how sad her life is and about what a crazy bitch she turns into. That's so lazy and pathetic." Others agreed and even tried to petition Netflix to ax the show. Unfortunately, the streaming network, where I was once employed, overlooked those protests and renewed the show for a second season. The show's cocreator and executive producer, Lauren Gussis, tried to quell the backlash by claiming that the plot was based on her own experiences and was a "cautionary tale about how damaging it can be to believe the outsides are more important—to judge without going deeper."

Isn't it beyond time we go deeper, though? Kelli, one of the central protagonists on HBO's half-hour comedy *Insecure*, is supposed to break the mold: She's a successful accountant. She doesn't exist solely to boost her thinner friends' self-esteem. She's funny as hell, forward about her expectations and standards, and doesn't tolerate mistreatment. Kelli is a modern woman, but when it comes to relationships, she's still depicted as desperate. In one episode, Kelli is shown thirstily lusting after a man in a club, and when she and her friends venture to the Waffle House, he fingers her under the table—as her friends look on uncomfortably and exchange knowing glances—and she falls all over herself to keep him interested. Kelli also pretends to be British in the season 4 finale—for no other reason than to impress the new man she's dating. While Kelli is a new kind of fat protagonist in many respects, there's still a lingering specter of the desperate fat Black woman hanging over her character. When Kelli interacts with men, her friends' reactions mirror how we, the audience, are supposed to view her: she's laughable, pitiable, deluded. Kelli gets her happy ending: In the show's final episode, she begins running the estate division at a Black-owned legal firm. She falls in love with a man who simply says he's going to stick beside her no matter what. She even reverses course about having children; during her birthday dinner, she announces that she's pregnant. It's a hard-earned ending. After all, it only took her alma mater accidentally declaring her deceased in her reunion brochure to give her a new lease on life and propel her into this new stage. Kelli's journey feels like a flashing red light—a reminder of how far I've come.

Dorian and I met on a dating site when I was in college, and we quickly connected. When I texted him, he responded. When I called him, he responded. When I attempted to make plans for us to meet, he went deathly silent. After three months of simply being pen pals, he finally agreed to hang out with me—take me to his house, fuck, the usual—and when I asked him why it had taken him so long, he deflected, saying, "I've been busy, but where am I on my day off? With you." Without asking, he was asking: Isn't that enough? We had mediocre sex. I faked an orgasm. And then we fell into that same routine. It took more than six months for me to move on, because it felt as if his attention was the center of my world. I never thought I could do better, because I rarely saw a fat Black woman modeling that reality for me.

Some might argue that there's value in having a fat protagonist on television, and that their presence should supersede any presumption of their impact. But aren't we past the point of representation being enough? At what point can fat people demand more fleshed-out characters? Because the absence of images makes underrepresented groups pine for scraps, fat people are encouraged to gobble up whatever representation is offered to us, no matter how shallow and full of sadness. For some, the mere presence of nonthin bodies in pop culture is worthy of breaking out balloons and confetti.

Take, for instance, the ongoing conversation about Lena Dunham's "radical" choice to showcase her nude body on prime-time network television. When *Girls* debuted on HBO in 2012, "real" was one of the words repeatedly used to describe the comedic drama about four twentysomething friends trying to figure out their friendships, their romantic relationships, and their careers in a gentrified area of Brooklyn where Black and brown people have either been pushed out by rising rents or transported to another planet. Though some critics have rightfully panned the show's clumsy handling of race, others have praised Dunham and her co-showrunner, Jenni Konner, for showing young (white) women of various shapes and sizes having sweaty and unglamorous sex, taking showers, and just being naked—as so many of us are when we're in the comfort of our homes.

As writer Ragen Chastain explained in an article for *espnW*, when "thinness has almost always seemed like a prerequisite," seeing normal women partaking in normal activities can seem revolutionary. Dunham's fully nude sex scene with Adam Driver in the pilot episode "stirred something in many women viewers," Chastain wrote. "For probably the first time, they saw someone who looked like them on the screen as a fully developed character—a sexual being, desired. That quickly became something that not only intrigued viewers but empowered and compelled them to tune in every week. Because representation, especially when you haven't had it before, quickly becomes something you want to experience as often as possible."

Chastain isn't wrong. When underrepresented communities see an image that's even remotely representative or familiar, it becomes an opportunity to both increase visibility and foster conversations about how the absence came to be. In multiple interviews, Dunham has said that Hannah, the character she plays, is often naked because she wanted to challenge the ways in which television has historically promoted thinness as the pinnacle of beauty in an unhealthy way that shuts out other bodies. Real and raw sex isn't a new phenomenon on television, though it was when *Sex and the City* debuted on HBO in 1998. Carrie (Sarah Jessica Parker), Samantha (Kim Cattrall), Charlotte (Kristin Davis), and Miranda (Cynthia Nixon), four thirtysomething professionals working, living, and fucking in Manhattan, changed the game from the series's very first episode, when newspaper columnist Carrie asks a seemingly simple question: Can women have sex like men? For six seasons, each woman answered that question in her own way. *Sex and the City* ushered in a generation of TV shows in which thin, white upper-class women caved to their carnal desires without regret or consequences, but that freedom was rarely afforded to those who fell outside of what's perceived as "normal." *Girls*, however, conveyed Dunham and Konner's commitment to normalizing different, albeit white, bodies.

In a season 5 episode of *Girls*, Hannah has a casual weekend fling with a gorgeous doctor. When the episode aired, some men, including *Slate* writer Daniel Engber, couldn't understand how she—a woman who some would consider "average"—could lure such a man. (A doctor!) "Jenni Konner has said that this season is about seeing what happens when Hannah starts 'getting some of what she's been pining for,' so I guess this was their attempt to wrestle with that theme," he said in a roundtable conversation. "But presumably there are things that Hannah would not, in any world that resembled our own, get. Such as [the doctor] for instance. I *want* to suspend my disbelief—just as viewers have, for generations, imagined that Al could get Peggy and Homer could get Marge and Jim Orenthal could snag Cheryl Mabel. But the show needs to work harder to make that seem feasible. And not pile implausibility upon implausibility."

For decades, our society has accepted without issue the slovenly male characters with bombshell wives that Engber cites. But being able to suspend disbelief is a right reserved for men, specifically white men, who perceive women's bodies as some sort of conquest. Male viewers like Engber happily buy into the fantasy that men with little more to offer than some charm will attract a conventionally beautiful woman. And yet, those viewers believe that Hannah's wit and openness should repel, instead of draw, a man. If average-looking men can have reciprocal relationships with gorgeous women on-screen, then why should we believe that women who exist outside of a narrow definition of beauty can't have lots of sex with all kinds of people who find their bodies attractive? But it's much easier for Engber to perceive Hannah's casual relationship with Wilson as an impossible feat than to explore how fatphobia and sexism dictate the kinds of relationships that are depicted on-screen.

Women celebrities of all sizes are treated as objects rated on a fictional "fuckable or not" scale. The more fuckable women celebrities are perceived to be by mediocre cis-het men, the more likely these women are to evade criticism. These delusional men aren't even within the orbit of these powerful, wealthy women, but they've persuaded themselves that they have a chance. In fact, these women would be lucky to fuck them. If said woman celebrity is Black or fat, this scale becomes even more restrictive. Women have to be thick but not fat. We can have wide hips but not stomachs that spill over our jeans. We have to be lighter than a paper bag. If our hair is natural, it has to be "neat." These delusions give men, like Engber, permission to judge what a nonthin woman is worthy of in her romantic partnerships. In his mind, it's surely not a doctor. The "ugly guy, hot wife" trope, wherein the woman is considered far more attractive than her significant other, supports a sexist fantasy that any man can get any woman. That's one of the consequences of having such limiting narratives about the experience of being fat, or just seeing bodies that are bigger than those of the thin actresses who populate most TV shows and movies.

I FOUND A LOT OF SOLACE IN TELEVISION AS A TEENAGER, ESPECIALLY when I was in the throes of one of my agoraphobic episodes. When my parents left for work and my brother left for school, I'd lie in the bed or on the couch and escape into TV shows for hours, living vicariously through guests on daytime talk and courtroom shows, reruns of classic 1990s sitcoms, and procedural dramas about detectives solving seemingly impossible crimes. While television gave me comfort, there still weren't many women on-screen with bodies like mine, so when I found them—on old VHS tapes or DVDs—I held on to them for dear life. I learned so much about myself, about my body, and about what awaited me from the fat women who graced my television screen.

In 2014, Laverne Cox and Janet Mock introduced the concept of possibility models into our cultural lexicon. Instead of a role model, or a person who's assumed to be devoid of flaws, possibility models are guiding lights that show us how we can be in the world. They are less definitive. We choose what we learn from our possibility models, while we project our own expectations of perfection onto role models. These are the possibility models who guided me through rough teenage terrain and helped me decide the kind of fat woman I wanted to be:

KHADIJAH JAMES ON FOX'S *LIVING SINGLE*

I will never forget the first time I laid eyes on Khadijah James, one of the four protagonists of Fox's hit '90s sitcom *Living Single*. More than twenty years later, that moment is still as resonant as being swept up in the magical world of *Matilda* for the first time or escaping into the foreign, but somehow familiar, world of *Gullah, Gullah Island* every day after school. At the time, I didn't have the right language to explain why Khadijah, played by the incomparable Queen Latifah, made me dazzle with happiness and a sense that I'd finally been *truly* seen. However, I did know, almost intuitively, that seeing a three-dimensional plus-size Black woman on television was transfixing, essential, and transformative. It helped that Khadijah's body—tall, sturdy, and definitely not thin—so closely resembled the adult body I imagined I'd have.

Khadijah helped me see a pathway forward, even as doctors, teachers, and other media imagery reminded me that happiness was guaranteed only for those who met a certain body mass index (BMI). By the time I discovered Khadijah, I was a teenager who escaped into chat rooms instead of facing a world in which fatness is associated with laziness, sadness, and, most important, loneliness. I'd really convinced myself that I'd always be alone because I was unworthy of companionship. Doctors reinforced my internal dialogue. Classmates reinforced it. Boys who called me "a fat bitch" when I didn't want to give them my phone number reminded me about it. But Khadijah, a confident and very fly magazine owner, made me think about fatness, and thus myself, differently.

James might be an exception to the rule, but she never realized it: she emanated a supreme confidence that didn't require validation from anyone, including her roommates and her romantic partners. Though she's clearly fat, Khadijah never discusses her weight. She owns her own magazine (an urban publication called *Flavor*, focusing on the New York hip-hop scene), supervises her employees effectively without compromising her vision, nurtures others as much as she nurtures herself, and dates a lot, even though she tends to create boundaries that allow her to maintain her independence. When *Flavor* hits a financial hiccup, she resists taking a $2,000 business loan from her best friend and neighbor, Maxine Shaw (played by Erika Alexander). That's because, as her childhood friend and roommate Regine (Kim Fields) points out, she was born uptight. "The doctor slapped her, and she slapped him back."

Khadijah carries that unshakable sense of independence into her romantic endeavors as well. She prioritizes her own desires, and she encourages her friends to do the same. She's the driving force behind persuading her closest friends to go to a nightclub to "mack" on men because "I need a real man, and real men are not going to just knock on our doors." When they arrive at the nightclub, Khadijah has no qualms about snagging a cutie, the club's DJ, and even belts out a tune at his behest. When Khadijah and Maxine compete for the affections of their new neighbor Hamilton (Morris Chestnut), Khadijah challenges Hamilton to a game of basketball in their brownstone's hallway. Khadijah isn't interested in being demure or allowing him to win—and her competitive streak is an asset in its own right; Hamilton chooses to take her on a date.

Our fatphobic society encourages fat women to place a lot of significance on being chosen, meaning we should settle for whomever offers us affection. But Khadijah refuses this mandate. The season 1 finale finds Khadijah ensnared in a love triangle with her childhood friend Scooter (Cress Williams), who moves from Los Angeles to Brooklyn to take a new job at a record label and rekindle their romance, and Alonzo (Adam Lazarre-White), a handsome and accomplished teacher who wants Khadijah to move in with him. Khadijah chooses Scooter—on the condition that he live and work in New York. When Scooter decides to embark on a world tour with a group that he's managing, Khadijah decides to end their relationship rather than ask him to give up his dream. "I love you, and I'm letting you go" is an empowering message—one that I wish I'd heard more often as I navigated dating. Instead, I settled—time and time again—because I'd internalized the idea that I wasn't worthy of a partner who valued and loved me.

By the time I transferred to Bennett College, a historically Black college for women in North Carolina, I had dated my fair share of undeserving men who always prioritized their pleasure at my own expense. I met Nicholas when he held the convenience store's glass door and opened his palm, signaling that I could walk through the entrance before him. I smiled, said thank you, and walked past, immediately sensing him surveying my entire body. He went into lockstep beside me and initiated a conversation about my plans for the weekend. I gave Nicholas my phone number, and in less than three hours, we were sweating in a dark nightclub, my ass bouncing on his lap as he gripped my hips and grinded into me. When he drunkenly whispered in my ear that he wanted to get naked, I grabbed him by his hand, walked with him outside, and let him drive me to his apartment.

From the moment he started lifting my shirt as we walked through his front door, I knew I'd made a mistake. I wanted to say no, to walk away, to go back to my college campus, but I didn't want to disappoint the person who'd fed me and filled me with alcohol. I had sex with Nicholas that night—and it was terrible. He touched me too roughly, kissed me so sloppily that spit dripped down my chin, and selfishly humped me fast and rough. When he came in the condom and got up to go to the bathroom, I felt tears building in the creases of my eyes. "I shouldn't have had sex with him," I told myself. "Why did I do that?"

By the time he came out of the bathroom and I went in, I'd pushed those tears and my discontent down. I was spending the night with a cute guy who got pleasure from my body and made sleepy promises about cooking breakfast in the morning. I sidled up close to him, let him loop his arm around my shoulders, and fell asleep on his chest. I had sex with Nicholas the next morning after we ate breakfast. We had sex later that afternoon, after he'd taken me to campus to shower, change, and get my laptop. We had sex that night, when I again spent the night with him. The sex never got better, but for nearly two months, I endured it—telling myself that bad sex was better than sleeping alone.

Settling for bad sex isn't something that just fat women do; many women fake orgasms to appease their partners and stroke their egos. When you're fat, however, there's an added pressure of knowing that many people, especially cisgender men, will publicly proclaim their disgust for fat people while willingly bedding them in private. I'd read that line enough on dating sites and heard it enough from men to believe that it was true, so I'd bought into the idea that lying down with a dude—even if it was unsatisfying—was an experience I should be grateful for. I tolerated Nicholas because I didn't think I deserved better than a selfish sex partner who had no idea how to make me orgasm and wasn't interested in learning. Khadijah was the first character I'd seen who told me I didn't have to just accept what was offered. Khadijah had options, and so did I. I unceremoniously dumped Nicholas after he promised me that he was picking me up for dinner and then never showed up. I wasn't sure if I could ever do better, but I didn't want to settle for a man who broke promises *and* couldn't fuck.

JANE BINGUM ON *DROP DEAD DIVA*

When *Drop Dead Diva* debuted on Lifetime in 2009, it was something different within a vast array of tragic sameness: The hour-long comedic drama centers around Deb (Brooke D'Orsay), an aspiring model who's killed in a car crash. When she arrives in heaven, she accidentally stumbles on a forbidden portal that transports her back to Earth in a different person's body. She returns as Jane (Brooke Elliott), a plus-size humdrum, drab, all-work-and-no-play attorney whose entire wardrobe consists of black suits. Over six seasons, Jane reinvigorates her closet, falls in and out of love, and wins massive cases for her fledgling law firm. Margaritas and cookies seem to solve all of Jane's woes, and she remains committed to her own happiness. Jane's life isn't undergirded by tragedy (other than her body being inhabited by the spirit of a dead person).

Despite the strangeness and offensiveness of a model having to enter a woman of size's body to give her spunk, Jane is the polar opposite of the usual stereotypes associated with fat female characters. She's beautiful and brilliant, and she knows it. She treats her body as a larger canvas, adorning it with clothing that accentuates her assets rather than camouflaging her perceived flaws. Jane understands that she disrupts spaces with her perkiness and her size, so she does it intentionally. She is the best friend we all want to have but never seem to find—especially on television.

Jane isn't invisible or a punch line; she's confident, sexual, and desired. Deb's fiancé, Grayson (Jackson Hurst), is the sexiest man in the law office. Though he was in love with Deb, a slender blond model, he finds himself falling for Jane. Eventually, Jane becomes engaged to Owen (Lex Medlin), a judge in front of whom she often argues cases. Owen is attracted to Jane's beauty and her intellect and never seems bothered by her weight. It's never mentioned when they're out to dinner. And he often stands back and admires Jane as she walks past him. She's quite literally the object of his affection, a fact that shatters the idea that fat women are undeserving of romantic love.

When *Drop Dead Diva*'s final episode aired on June 22, 2014, it represented the end of an important era in fat representation on-screen. There were no other series that could rival the work the show had done in reimagining the fat female protagonist. Josh Berman, *Drop Dead Diva*'s openly gay creator and executive producer, wanted a show that honored his grandmother Deb, a "4-foot-11, pudgy Holocaust survivor who carried herself like she was a supermodel." She was Berman's hero and his role model because she refused to allow life circumstances to dictate her happiness. "That was the genesis of *Drop Dead Diva*," he told *Princeton Alumni Weekly* in 2010. "How are your grandmother and the character Deb in Jane's body similar?" With his grandma in mind, Berman infused specificity into the character and into the show. Real-life Deb was confident, so fictional Deb is as well. "The character doesn't look the part and neither did my grandmother," Berman said—and that was the point. He depicted life as it really is for people on the margins of society. The show won two GLAAD Media Awards for hiring trans actors to portray trans characters and a number of queer icons guest-starred or made cameos during the show's run. Berman said in an interview with *USA Today* that in Hollywood, "beauty has been defined as size 2 and under 25; hopefully we can help redefine the paradigm." *Drop Dead Diva* did exactly that.

BETTY SUAREZ ON *UGLY BETTY*

I've been in a semiabusive relationship with the fashion industry since the moment I exited the womb. I fell hard for it when I was seventeen, after engaging in a valiant war against that love for more than a decade. After all, I was a chunkierthan-most kiddo growing up in the Kmart "polka-dot era" long before prints were fashionable. I dreaded going shopping; it was a chore to browse through racks and stacks of blouses and find shirts that weren't even cute enough to sleep in. But at the age of seventeen, after having my final agoraphobic episode, I decided that I needed to define style for myself, even if that meant digging deeper into clothes bins to put outfits together. I owe that life-changing epiphany to *Ugly Betty*, an ABC dramedy about Betty Suarez (America Ferrera), a twenty-two-year-old Mexican American from Queens, New York, who's thrust into the cutthroat fashion world when she's hired as the assistant to Daniel Meade (Eric Mabius), the editor in chief of *Mode* magazine.

From 2006 until 2010, I watched Betty navigate the infamous fashion world in hideous knee-highs and oxfords. Despite numerous missteps, she still managed to achieve some semblance of success. She leveraged her assistant position into an associate features editor spot at *Mode* before being whisked to England to assist with an upstart magazine that promoted women's empowerment. Betty chased her dreams, and I closely followed her journey. I felt a kinship with Betty as she balanced her burgeoning career with supporting her undocumented father, queer nephew, and sister who's a single mother.

Betty evolved from a television character into my personal hero. She was from Queens, worked at an elite fashion magazine, had a normal body, and wore braces. Betty was spunky and ambitious but also had firm morals and boundaries; she refused to compromise her values to climb up the editorial ladder. I was enthralled. All she toted from Queens to *Mode* was a pen and a dream. I'd traded in traditional paper and ink for a laptop and Microsoft Word, but the principle was the same. Midway through the first season, I realized that, like Betty, I had to endure what the world threw at me. I loved pairing weird pieces, and I researched Diane von Furstenberg and Anna Wintour like philosophers google Socrates; but when I flipped through these magazines, most of the women featured in editorial spreads and advertisements bore no resemblance to me. These ladies were tall, rail thin, and white, with straight blond hair, piercing blue eyes, and photoshopped-to-perfection skin. The only thing I had in common with these "stunning" women was my "bone structure," a pickup line that creepy photographers in New York's LaGuardia Airport have attempted to use on me several times. I felt like an outsider. I wasn't confident that I could infiltrate fashion or effect change as Betty did on television.

Then, I got invited to cover New York Fashion Week (NYFW) in February 2012. I was a senior in college, rubbing elbows with fashion's elite, sitting in the same rooms as editors in chief and trendsetters. On my first day in New York, however, my excitement waned—quickly. As waiflike model after paper-thin model walked down the staircase at the Rubin Museum of Art, I realized that I was the sole large woman in the room. At first I didn't mind; after all, I knew that Monif C. and other popular full-figured clothing designers weren't unveiling fall 2012 collections this Fashion Week. And when I stepped off the elevator backstage at Levi's, I still felt a thrilling jolt. But not seeing models, journalists, celebrities, or other attendees who resembled me in any way, from Carmelita Couture to bebeBLACK, was disheartening. I love seeing large, confident, and beautiful women grace the catwalk—and, disappointingly, this void was not filled. Even Toccara Jones, then the go-to plus-size model, was absent.

But this blatant exclusion of women with large hips and wide waists, who represent more women than frail super-models do, reflects the overall culture of the fashion industry. Many other NYFW attendees were cold to me; fashion is an exclusive industry, and I didn't have the right connections to secure a smile from even the lowest editor on the totem pole. Yes, I earned tickets and first-row seats to events just like the rest of them, but it didn't matter. It hurt like hell. According to a 2006 *USA Today* article, most runway models weigh an average of 120 to 124 pounds and are between sizes 2 and 4. Models weigh 23 percent less than the average female and 50 percent of models are underweight. It's no surprise that this industry once proclaimed supermodel Tyra Banks "too large."

Though these statistics are alarming, they haven't kept industry professionals, including People's Revolution owner Kelly Cutrone, from proclaiming that such standards are nec-essary for sustaining the fashion industry. "Women shouldn't be comparing themselves with these girls. These girls are anomalies of nature. They are freaks of nature," Cutrone ex-plained in an interview. "If we get a girl who is bigger than a 4, she is not going to fit the clothes. . . . Clothes look better on thin people. The fabric hangs better."

While attitudes about plus-size models have evolved since then, Cutrone's harsh perspective was entrenched for a very long time. So, no, I didn't come to Fashion Week expecting to see size-18 hips dominating the catwalk (though that would have been a welcome change of pace), but being the largest woman in the room forced me into the "sore thumb" position. After that NYFW, I stumbled on plus-size fashion blogs, such as Gabi Fresh, Gorgeous in Grey, LoveBrownSugar, and Girl with Curves, and discovered the explosion of "niche" clothing stores springing up across the United States and abroad. I was excited that the fashion world was expanding to include more women who resembled me. I found faith and resilience in these ladies. They were blazing trails. It was amazing to watch. I wanted in again.

In the final season of *Ugly Betty*, all these doors open for Betty. But she realizes that even though she loves *Mode* and has grown accustomed to the fashion world, it isn't the life that she wants. After that brisk week in February, I, too, evolved. While I don't professionally write about the ins and outs of fashion as I once did, I realize that everything I touch has flair and passion and individuality. I am fashion and nothing can ever take that from me. I owe so much about the plus-size woman I've become—unapologetically uncompromising about my values—to fat characters who exuded those traits on-screen when I was incredibly vulnerable. Or, as Jane says in an episode of *Drop Dead Diva*, "This doesn't have to make sense to you. This is my life. And I'm gonna live it the way I want."

ALL HAIL THE QUEENS OF
NOT GIVING TWO FUCKS

O ur bathroom's oval-shaped mirror is covered with streaks, flecks of toothpaste, and the dirt that comes from two children who create tornadoes of grime wherever they lay their hands. When I flip the light switch, the cylinder-shaped bulbs that surround the mirror light up, illuminating my preteen body and the large humps that have overtaken my torso. When I wear my favorite short-sleeve burgundy shirt, with its V-neck and soft, cashmerelike material, I am most aware of my titties. I'm ten, and I've already fully blossomed. Menstrual cycle, check. Hips, check. And breasts, definitely check. Gravity hasn't yet caused my breasts to droop, so they're perky and supple, and they're even more appealing when I pull down the shirt just right, so cleavage fills the V-neck.

Before I developed a grown-ish woman's body, I'd stand before this same mirror, disappointed that the buds on my chest hadn't sprouted in any noticeable way. I'd pull my shoulders back, trying to make my chest seem fuller, and then I'd stuff tiny balls of tissue in the front of my shirt, cupping my hands underneath my makeshift breasts to keep them from falling to the floor. But now that I have actual, real-deal breasts, I want to flaunt them as much as possible. Overnight, I've gone from being the butt of fat jokes at school to being an object of fascination, a maturing preteen to be fawned over. Instead of overlooking and ignoring me, boys start asking me for my home phone number so they can call me after school.

I want to maximize that feeling on my terms, so at least twice a week, I whip out that burgundy shirt—much to the chagrin of my parents, who aren't fond of my newfound womanliness. That shirt makes me feel sultry and attractive in a way that I feel as if I have control over, just like the Black women singers who dominate the music videos I play on an endless loop.

IN THE EARLY 2000S, HIP-HOP AND R&B MUSIC VIDEOS WITH BIG BUDGETS and perfect bodies aired over and over again, especially on popular countdown shows such as MTV's *Total Request Live* and BET's *106 & Park*. The spring and summer of 2002 and 2003, in particular, stand out in my memory because of the number of young Black women artists who dominated the charts and those coveted countdowns. In April 2002, Amerie released "Why Don't We Fall in Love," a perfect summer anthem about a woman declaring her intentions to a potential beau. While the song itself is only two minutes and thirty-nine seconds, the carefree video intensifies its impact. The video features Amerie leaving her apartment in a thin white T-shirt, mini-shorts, and red open-toed stilettos, and walking through her buzzing brownstone-lined New York neighborhood, stopping at a park, waving at neighbors as they soak in the warm weather, and just giving off the kind of vibe we all pine for in the summer.

In "Rock wit U (Awww Baby)," the lead single from Ashanti's sophomore album *Chapter II*, Ashanti sings about the early stages of love as she saunters through a tropical jungle, twisting her hips in front of an elephant submerged in enviable blue waters, and dancing seductively in a windowless bungalow.

Though Tamia offers a slightly different vibe in "Officially Missing You," a song about the grief of losing a romantic partner rather than newfound love, she still turns heads as she walks through a New York City neighborhood in wedges with laces that wrap around her lower legs. I vowed to become an adult and dress as she did. Then there is Mýa, contorting her lean body into all of kinds of impossible dance moves in "My Love Is Like . . . Wo," a Missy Elliott–produced track that convincingly sells the idea that the singer can change a partner's life. And, of course, Beyoncé dominated that summer from the moment she strutted onto an elevated wooden platform in a sleeveless white T-shirt, jean shorts, and glistening red pumps. "Crazy in Love" usurped the competition, creating iconic looks that I, for one, tried my absolute damnedest to emulate.

These sultry R&B divas introduced me to what I perceived

as womanhood—at least for the time they dominated my television screen. I learned about being sexy from watching Beyoncé tossing her voluminous blond curls in the "Naughty Girl" video, as men stopped lighting cigars to admire her. And from Aaliyah, body rolling in the Caribbean in a crop top, tight jean capris, and sneakers, while she sings about the sexual pleasure she hoped her partner would bestow on her.

While these R&B icons were helping guide young Black girls toward a semblance of self-realization about the beauty of our skin and our curves, there were a parallel group of women—not musical artists, but artists nonetheless—who were also influencing how so many Black girls and women thought about themselves and their bodies. An elite crop of video vixens, including Melyssa Ford, Buffie the Body, Vida Guerra, Karrine Steffans, and LoLa Monroe, often stole the spotlight from the rappers in whose videos they appeared. Those bodacious women oozed confidence as they danced alongside artists, and essentially functioned as legitimate costars in mini-movies that sometimes cost upward of $500,000. Video vixens were almost as famous as the artists with whom they worked, and they embodied a different kind of beauty ideal. Most of them weren't thin in the traditional sense, but their waists were small, and they had large breasts, wide hips, and round asses that I deeply envied.

Men could sit drinks on their asses—and the glasses wouldn't fall over. And in those forbidden, adult-only videos that came on at two in the morning on BET's *Uncut*, women with curves for days, small waists, large breasts, and wide hips drove men wild as they twerked in soapy bathtubs, on stages with stripper poles on which they effortlessly slid up and down, and even in front of rappers who disrespectfully slid credit cards down their ass cracks.

I wanted to be these women—with expensive clothes, curves in all the "appropriate" places, and men falling over themselves to have conversations with me, rub oil over my ass, and touch my body in sexual ways. When you're trapped in the house, with music videos, books, and internet chat rooms as your only entertainment, fantasies can become a form of survival. I longed to be desired, to be seen as worthy of attraction, to have someone stare at me with that glazed-over look of lust that crossed the faces of the men within these music videos and those watching them on repeat on television.

But based on the bodies presented as ideal in these music videos, I could never live up to the fantasy. My stomach wasn't flat enough. My upper arms weren't small enough. My waist-to-hip ratio wasn't the right proportion. My skin had too many scars, blemishes, and stretch marks to be displayed in bikinis that leave next to nothing covered. And, for some odd reason, I couldn't see my vagina. Of course, I knew what it looked like. Who hasn't taken a mirror, put it down to the vagina, and spread their lips open, just to see if it's pink or brown? But when I looked down, I could see my brown stomach, covered in glorious stretch marks, and my ten unpedicured toes, but I could never, ever see my vagina. In most ways, it didn't matter—who's looking at their coochie every day or even thinking about doing that? But, for me, this deficit separated me from the video vixens and R&B powerhouses

I admired. A Coca-Cola-bottle frame, flat stomach included, was the marker of being able to star in a music video, be on television, and twerk so hard that men felt compelled to throw hundreds of dollar bills at you.

Missy Elliott, a pioneering triple threat who writes, produces, and creates mind-blowing music and videos, was the one woman of this era whose body resembled mine. Though she'd been on a very public journey of losing, gaining, and losing weight, Missy was the consummate big girl with wicked creative chops. I'd first seen her sporting black finger waves in the 1997 video for "The Rain (Supa Dupa Fly)," where she dances in a large black trash bag, fly two-piece yellow suit, and other colorful outfits, and just as when I saw Queen Latifah for the first time, seeing her increased my inner understanding of possibilities. Here was a dark-skinned, fat woman dancing effortlessly and damn near creating her own language—have you ever tried to sing the lyrics of "Work It" without messing it all up? Yeah, me neither.

Missy's music is inherently sexual. "Get Your Freak On" is one of the kinkiest songs of all time, and "One Minute Man" isn't that far behind. She's vocal about desiring a "hot boy," and "Work It" literally has the lyric "Love the way my butt go bum-bump-bum-bump-bump / Keep your eyes on my bum-bump-bum-bump-bump." Yet, she's never been positioned as the sex symbol she so clearly is. Instead, her appeal is attributed solely to her creativity and ability to innovate—her music videos are surely out of this world, but so is her proclamation that fat women deserve healthy relationships full of good sex that lasts longer than a minute. In an ideal world, Missy would've been the woman I wanted to model myself after, but she was never considered "raunchy" or even beautiful. Instead, she was the asexual fat friend who could make people laugh and make them think, not the woman who gets chosen at the end of the video and swept into a luxurious life full of furs, jewels, and expensive trips.

There's a social presumption that Black women are nurturing, the kind of people who sacrifice themselves to take care of others. This expectation is compounded by race and by size, and it begins where everything for Black people's racial oppression begins: slavery. The role of the Mammy traces its origins to the enslaved Black women who were forced into domestic servitude for the benefit of white women: caring for their children, cleaning their homes, cooking for their families, and more. Cruelly, they spent more time caring for white children—curing them, nursing them, being substitute mothers—than with their own children. But in a Confederate effort to make it appear as if Black people were content with being enslaved, the Mammy became a dominant archetype in the U.S. cultural imagination.

For those who cared only about profit and insisted that slavery was a "necessary and kind institution," Mammies were essential to presenting the institution of slavery as something Black women not only approved of but wanted to remain as the dominant economic system in the United States. White people were so obsessed with this false image, in fact, that congressional leaders attempted to have a sculpture of a

Mammy erected in Washington, D.C., in 1923—as an ode? a way of misremembering history? a way of continuing to exploit Black women? All of the above.

Thankfully, such a memorial was never built, but the Mammy still resounds, particularly in the pop culture we consume. This archetype is entrenched in the way we view Black women, and fat Black women in particular. An essential part of the Mammy figure is her fatness. The Mammy figure is never sexualized or considered beautiful, even in a culture where Black women are so commonly objectified and hypersexualized. For fat Black women now, the ghost of that stereotype still permeates. It's the reason I didn't see Missy as an ideal—even as a child, I knew that that dark-skinned fat woman couldn't be idolized and thrust on a pedestal to be admired in the way video vixens were. Missy was supposed to strictly entertain us, and though we loved her in return, that love was never romantic or sexual.

Video vixens helped create a fantasy that I so desperately wanted to emulate, and instead of questioning how those women came to be idolized in the first place, I internalized a feeling of unworthiness. I couldn't live up to their standards, and, therefore, I would never be desirable—except, of course, when I wore that burgundy T-shirt.

As the music industry's profits declined, so, too, did the budgets for music videos. Many urban models of the 2000s found jobs serving unreasonably expensive bottles of alcohol at popular hip-hop hangouts or showing off their still curvy and flawless bodies on Instagram. Some are now paid thousands, and sometimes hundreds of thousands, of dollars to advertise clothing lines and laxative teas. No matter how much time has passed, they still seem to maintain their good looks.

Urban models such as Lira Mercer (formerly known as Lira Galore), Bernice Burgos, and Shannon and Shannade—known as the Clermont Twins—are often referred to as the "Black Barbies of Instagram." They've accrued millions of devoted followers on social media who track their every move, admiring the sculpted bodies they flaunt in shoestring bathing suits. They're the newest iteration of vixen, a new ideal to aspire to with their unreal hip-to-waist ratio and ability to strike the perfect pose. In many ways, these flawless women appear to offer up an enviable lifestyle; they're constantly traveling via private jet and enjoying the fruits of luxury. They then use Instagram to promote their own clothing boutiques, makeup lines, detox juices, waist trainers, and a host of other products that feed on women's insecurities.

Their photos and captions imply that these products will bestow ordinary women with the bodies, wealth, and opportunities they have—and women the world over believe them. Yet, these Instagram influencers rarely reveal the names of their surgeons, the procedures they've had to make their sculpted bodies appear natural, and the thousands of dollars they've spent on disappearing waistlines and slim nostrils. There's nothing at all wrong with women correcting their bodies as they see fit, and not all these models have had plastic surgery. But there's no doubt that many rely on more expensive and invasive means than simply contouring their noses and wearing waist trainers to achieve their looks. Pretending that their bodies came through hard work and diet just creates a never-ending cycle of dysmorphia and dysfunction for the women who follow them.

In a world where the majority of us are connecting via social media, Black Barbie influencers feel inescapable. While we can all scroll past these ads or stop following the people shilling them, we can't leave a society that thinks waist trainers and detox teas are safe. More than 500 million people use Instagram every day, and the platform has 1 billion active users each month. Similarly, Snapchat has more than 300 million active users each month, and more than 188 million users send Snaps every day. Snapchat and Instagram market filters that alter the user's face—whether by lightening skin, adding dog ears and flower crowns, or subtly enlarging the eyes and mouth—as perks of using the platforms.

While social media is in some ways a revolutionary tool for positive change, researchers at the Boston University School of Medicine have found a direct link between using Instagram and Snapchat filters and the development of body dysmorphic disorder, which is "characterized by persistent and intrusive preoccupations with an imagined or slight defect in one's appearance," according to the Anxiety and Depression Association of America. Those researchers have named this troubling phenomenon "Snapchat dysmorphia," which they further detailed in an August 2018 article published in *JAMA Facial Plastic Surgery.*

The authors of the *JAMA* article said that Snapchat, Instagram, and Facetune—an app used to photoshop images—are "altering people's perception of beauty worldwide." That has dangerous implications. Body dysmorphic disorder (BDD) "is more than an insecurity or a lack of confidence," the study concluded. "Those with BDD often go to great lengths to hide their imperfections . . . and may visit dermatologists or plastic surgeons frequently, hoping to change their appearance."

"There's an issue with losing perspective on what you actually look like, and it's not something we talk about much," Renee Engeln, professor of psychology at Northwestern University and author of *Beauty Sick: How the Cultural Obsession with Appearance Hurts Girls and Women*, explained to *HuffPost* in 2018. "It's not enough [to] have to compare yourself to these perfected images of models, but now you've got this daily comparison of your real self to this intentional or unintentional fake self that you present on social media. It's just one more way to feel like your [sic] falling short every day. It's a real bind we put women in when we give them this non-stop pressure to conform to this particular beauty ideal and then try to shame them when they feel that pressure and they do something about it. I don't think that's the key here. I think what we want to be moving toward is more real representations of ourselves and I think that's becoming harder and harder."

After my mom bought a waist trainer she'd learned about on Instagram, she asked me to help her put it on. She stood shirtless in the center of the kitchen, one hand holding the center of the trainer against the middle of her stomach, as I walked around her, pulling the elastic band tighter and tighter up her abdomen. When I closed the waist trainer, she couldn't breathe; it was almost if she were wearing a corset. But all she could worry herself with was the trainer not fitting as well as the one that fit the Instagram model. None of these products can magically transform an ordinary woman into the celebrity women they're admiring.

Actor Jameela Jamil has been campaigning on social media to end this practice of celebrities shilling ineffective products for the last few years. When Cardi B began promoting detox tea, for instance, *The Good Place* actress called her out: "They got Cardi B on the laxative nonsense 'detox' tea," she tweeted. "[God] I hope all these celebrities all shit their pants in public, the way the poor women who buy this nonsense upon their recommendation do." SkinnyMint, SkinnyMe, Flat Tummy Tea, Slendertoxtea, Bootea, Skinny Teatox, and other celebrity-endorsed and -hawked detox teas are not only inefficient but also incredibly unhealthy. Ordinarily, detox tea programs include both a daytime and nighttime tea that contain a host of different caffeine stimulants, including guarana, green tea, and yerba maté, that are designed to increase metabolism.

Using laxatives, which is what many of these teas actually are, can cause serious long-term harm. According to the National Eating Disorders Association, the health consequences range from electrolyte and mineral imbalances to abnormal bowel movements and permanent colon damage that causes infection, and, in rare cases, colon cancer. And, for those with high blood pressure or heart issues, a medicine-induced increase in metabolism can be particularly dangerous. Yet, Instagram in particular has allowed laxative teas, lollipops, and other products to be sold on its platform with minimal regulation because the U.S. Food and Drug Administration (FDA) hasn't issued guidelines for how these products should be used or if they're safe to use at all. "The FDA does not regulate these substances," Dina Halegoua-De Marzio, a gastroenterologist, told CNN in 2018. "Really, they can put anything they want in these teas. They can make any claim that they want, and it doesn't have to be supported by any evidence. That makes these very, very dangerous."

Years after video vixens were popular, a new generation of girls and women, especially girls and women of color, are dissatisfied with their bodies because they can't achieve what's being marketed as ordinary. Former child actor Raven-Symoné has been incredibly public about her complex relationship with her body. During her career, which began on *The Cosby Show* and was followed by her own Disney Channel show, Symoné's body has been heavily scrutinized. In a 2017 interview with *People*, she said incessant body-shaming throughout her career caused her to develop mental issues. She wasn't allowed to eat on the set of *The Cosby Show*, and people questioned whether she should be cast in a Cheetah Girls movie because of her size. "It was definitely hard," she said. "I remember not being able to have the bagel or anything at—we would call it crafty, where it's just a table of food, ready for you to eat whatever you want. And I remember people would be like, 'You can't eat that. You're getting fat!' I'm like, 'I'm seven! I'm hungry!'" Symoné ended that interview saying that "the world is too big to have one sort of view to show beauty, because then you are literally destroying society."

Yet, in 2021, she appeared on *Good Morning America* to share how she'd lost thirty pounds in three months by keeping a low-carb diet and being an "avid faster," someone who doesn't eat for a minimum of fourteen hours between dinner and breakfast and sometimes undergoes multiday fasts. Instead of eating, as the body requires, she drinks a lot of water and electrolytes and has a little bone broth when needed. "Every time I lost weight in the past, it was about size," she said. "[Now] I want to make sure that my body is healthy and prepared to deal with old age."

Good Morning America seemed to have no reservations about airing an interview that glorified subsisting on water for days at a time—and the fact that Symoné's fasting was presented in such a positive light reveals the pervasiveness of detoxing as a form of "healthy" and "acceptable" dieting. Other pop stars and cultural figures have promoted fasts and detoxes, even those who are otherwise known for body positivity. Even Lizzo, our culture's proudest fat pop star, has fallen victim to the idea that detoxing our bodies is a form of wellness that's unrelated to dieting.

Lizzo is inescapable: though she's been producing, writing, and performing music for more than a decade now, she's been on an unstoppable run since the 2016 release of her EP, *Coconut Oil*, and the subsequent release of "Truth Hurts." It's a perfect song that starts with a now iconic line: "I just took a DNA test / Turns out I'm one hundred percent that bitch." Lizzo's most successful single slowly simmered for nearly two years before finally being recognized as the misandrist gem it is and shooting to the top of the Billboard Hot 100, simultaneously catapulting Lizzo into the pop stratosphere. Since then, Lizzo's music has become as infectious as her personality appears to be. Her songs appear in commercials on a loop, show up in movie trailers for all kinds of films, and are in regular rotation at Top 40 stations across the country.

There's nothing unusual about Lizzo's music or her brand packaging. She's a classic pop star. She easily traverses musical genres—pop, R&B, and hip-hop—because she's tapped into the magical formula: her songs are catchy and absolutely unforgettable. Like other artists of this time, Lizzo's also mastered social media. Whether she's vacationing in a tropical paradise with a group of friends or twerking in a hotel room, she knows how to use her platforms to generate conversation and drive her followers toward her music.

Best of all, each of her albums is attached to an "era," a strategy reminiscent of what her pop predecessors, including Britney Spears, Christina Aguilera, Rihanna, and Beyoncé, have mastered—a new aesthetic to accompany new music. During her *Cuz I Love You* era, Lizzo appeared fully undressed on her album cover, looking directly at the camera as she sat on the ground with the soles of her feet pressed together. Of course, using nudity as a marketing strategy may be the oldest trick in the book. Following Janet Jackson's iconic 1993 *Rolling Stone* cover, which featured her topless with her then husband, René Elizondo Jr., covering her breasts with his hands, her album *janet.* debuted at the top of the Billboard 200. And it solidified the idea that she was fully independent from her family, with the most lucrative musical contract in the industry to boot. Madonna spent the entirety of her early career pushing the boundaries around women's sexuality, and though her photo shoots and videos generated some controversy, they also propelled her music sales. In other words, pop stars have used this tactic since time immemorial.

But for Lizzo, nudity—whether it comes via a curated Instagram grid or showing off her cheeks at a basketball game—becomes conversation for a simple reason. She's fat, unapologetically so, and she has no qualms about expressing how much she loves herself.

Lizzo is just attempting to create beautiful art and share the parts of herself that inspire her fans to live more authentically, but we project a whole lot of meaning onto her existence that she's never consented to or expressed interest in. Desperation is a defining characteristic for fat Black women, especially those in the public eye. They're desperate to lose weight and start their lives in a thinner body. They're desperate to fit in socially, making self-deprecating fat jokes at their own expense. They're desperate to be loved, even if said love is unhealthy. Lizzo doesn't appear to embody any of those characteristics, and we don't know what to do with her unapologetic love for herself.

Lizzo's 2019 album, *Cuz I Love You*, captures the ethos that runs through her music as much as, if not more than, her previous projects. She puts herself front and center as a woman deserving of self-love. It's a revolutionary idea, because fat people are asked to sacrifice so much to make others comfortable. Lizzo says fuck that, unapologetically encouraging other fat people to love themselves as she does. On "Soulmate," the fourth song on the album, she expresses that she's her own best thing: "Cause I'm my own soulmate (Yeah,

yeah) / I know how to love me (Love me)," she sings. "I don't need a crown to know that I'm a queen," she croons on "Scuse Me," a song from *Coconut Oil.*

Lizzo clearly loves herself, and yet, in 2020, she chronicled a ten-day detox on social media. During this detox, she drank tea and smoothies, ate apples and peanut butter, and popped supplements that were supposed to help her clean out her liver and start fresh. To top it all off, she posted a before-and-after photo to show off her progress. "I would normally be so afraid and ashamed to post things like this online because I feel as a fat girl, people just expect if you are doing some sort of thing for health, that you're doing it for some dramatic weight loss," she said in her Instagram Story. "That is not the case. In reality, November [2020] stressed me the fuck out. I drank a lot. I ate a lot of spicy things that fucked my stomach up and I wanted to reverse it and get my health back to where I was." Functioning livers detoxify human bodies, so there's no need to detox, but it has still been peddled as an easy, simple, and safe solution.

Lizzo understands the dangers and pitfalls of diet culture, but even she fell victim to the allure of being healthier—whatever that means—and the benefits of clearing out your system. Diet culture is so pervasive that none of us escape it, not a soul, and so we find the language to excuse our participation in it. As Lizzo proclaimed, the detox helped her sleep better and improved her mental health. "My fucking body, my fucking skin, my eyes, like I feel and look like a bad bitch," she said. "And I think like that's it. I'm a big girl who did a smoothie detox." What does it mean for the rest of us when one of the world's most visible fat people gets snared in the same trap that has been tripping us all up?

Lizzo exists in a unique space that many fat women find themselves in: since she's living a full life without concern for her weight, she has been labeled body positive and held to the standards of the movement without her consent. It's an experience common for fat folks who don't hide their bodies and actively reject diet culture. Rather than admiring us on an individual level, we become representative of an entire movement to dismantle fatphobia in every facet of our lives—and we're held to the standards and political commitments of that movement, making it impossible for us to exist without others concerning themselves about what we do with our bodies and how we do it. "The body-positive movement is the body-positive movement, and we high five. We're parallel. But my movement is my movement," Lizzo told *Allure* in 2019. "When all the dust has settled on the groundbreaking-ness, I'm going to still be doing this. I'm not going to suddenly change. I'm going to still be telling my life story through music. And if that's body positive to you, amen. That's feminist to you, amen. If that's pro-Black to you, amen. Because ma'am, I'm all of those things." In other words, Lizzo is simply attempting to exist—critics be damned.

Every fat woman doesn't get access to the same level of public attention, adoration, and revelry that Lizzo basks in. Though she's undoubtedly bombarded with fatphobia—to the point that she's posted her workouts online to debunk the idea that fat people are unhealthy and incapable of participating in sustained physical activity—she's also on the small side of the fat spectrum, which allows her to access everything from doctors to clothing in ways that super-fat people are unable to do. That doesn't mean, however, that Lizzo isn't constantly reshaping the ways in which we understand what it means to have a positive body image. Celebrities who peddle dangerous products, like detox teas and appetite-suppressing lollipops, are preying on populations of women who may have body insecurities but don't have healthy outlets to communicate their issues, even when they evolve into body dysmorphia or disordered eating.

Here's what I now know: I could never be a video model. I could never be an Instagram model. I may never see my vagina. And fighting to be in a world that normalizes bodies that look so different than mine is difficult, but there is no easy fix to such a systemic problem. No amount of Flat Tummy Tea can change these realities. But what can change them? Seeing women like Lizzo unapologetically living—not hiding, but living—can help us foster better self-images. So, put down the tea, stop following women who tell you that there's something inherently wrong with your body, and get comfortable in your own skin instead. That's the biggest lesson we can learn from Lizzo. It's all she's been singing about from the very beginning.

TURN OFF THE LIGHTS

The room was pitch-black with only a quarter moon creating shadows around our bodies through the slightly cracked blinds. Keith hovered over me, lowering the weight of his thighs between mine, as he smothered my lips with his, pressing his tongue into my mouth. My eyes were tightly closed, scrunched together, even, as I let the moment—my first sexual encounter—overtake me. As he rocked back on his knees, I leaned upward and clumsily tried to unbuckle his belt, while he removed his T-shirt and threw it over the side of the bed and onto the floor. Once he'd stood and removed his jeans and his gray boxers, he looked at me expectantly. It was my turn. That's when I froze—surprising myself and catching him off guard as well. Fluid pooled between my thighs, signaling a willingness to receive him, but undressing—unveiling my full body to a romantic partner—was beyond my scope of comfort.

Nudity, for me, has always been a natural state of being.

I'd strut through our apartment as a child in an undershirt, panties, and my mother's high heels so often that one of my uncles used to jokingly tell my parents and my grandmother that I was naturally inclined to stripping. More than twenty years later, I still love being naked. Is there a better feeling than coming home after a long day, stripping off stifling clothes, bra, and shoes, and finally breathing a sigh of relief as a breeze roils across your body? It's one of the things I dream about in the middle of the day, as my pants start to feel constricting and my bra digs in, irritating my shoulders and the sensitive skin under my breasts. Most of my female family members have seen me naked and pestered me with questions about how to be more comfortable and confident about their own nakedness. My cousin and I had never shared a room before we embarked on a cruise to the Bahamas in 2014, so we tiptoed around the delicate dance of creating and enforcing boundaries. She'd come out of the shower with a towel, sliding on a shirt before lotioning her body. I, on the other hand, walked around the room carefree and without a worry about how she'd perceive my body. "How are you so comfortable being naked?" she asked the night before the trip ended. I shrugged it off, not remembering that I'd once hidden my body from the man I was fucking.

I lost my virginity while wearing a white T-shirt. Each time Keith placed his warm palms under my T-shirt to grip my hanging breasts or lightly pinch my nipples, I'd flinch, worried that his erection would go flaccid the minute his hand made contact with my stomach, where a mixture of light and dark stretch marks crisscross my belly like train tracks. It was an irrational fear that didn't reflect the reality of our burgeoning relationship. My fear of rejection prevented me from acknowledging the passion or the care with which he touched me, as he paid special attention to the places that made me gasp in ecstasy. I'd internalized the idea that I wasn't worthy of his attention, his attraction, and his admiration. I wasn't prepared for how social messaging about the attractiveness of fat bodies would invade my first sexual encounters.

I'd been learning the grooves of my body since I first touched my brown vagina at the age of fourteen and realized that rubbing my clitoris at the right angle with just enough pressure would cause my vagina to tighten and then convulse, ushering in the most pleasant feeling I'd ever experienced. Every night, I'd lie under the covers of my full-size bed, spread my legs comfortably apart, and rub my clitoris until I orgasmed, breathing heavily and biting my lip to keep my sweet nothings from floating through the wall and awakening my parents.

Self-pleasure came naturally, and it became a salve for the loneliness that plagued me as I delved deeper into an online romance with a person who had no intention of bringing our relationship offline. Instead, I settled for dreaming of the moment when I'd be able to share that moment of climax with someone else. When Keith and I began communicating through Facebook a few months before I started college, I knew pretty quickly that he could possibly be that person. I would be enrolling in college in January, a full semester after everyone else, because I needed to save money and prepare myself mentally and emotionally to move across the country and away from my safety net—my parents, my therapists, and the friends who breathed life into me when death seemed to be a better option than navigating relentless pain. Knowing how large of a leap I was making, I began building my village before I departed. I started befriending fellow freshmen on Facebook, including my soon-to-be roommate and our suitemates, the friends they tagged in their photos, and anybody who was also starting the collegiate experience.

About two weeks into this friending spree, I stumbled on Keith. He immediately sent me a message introducing himself and asking me why I'd decided on that college, what I was majoring in, what my hopes and dreams were, and, of course, my relationship status. As the day of my departure drew closer, we continued talking every day and within a week of my arrival on campus, we were meeting in person, sharing a meal in our school's cafeteria. Two months later, we were frolicking in his bed, kissing, and feeling the desire building between us, and I was frozen—incapable of removing the one remaining barrier between us. In that moment, that T-shirt was like a comfort blanket that made me feel protected at a time when I felt the most vulnerable. While this might not be true for all fat women, it was definitely true for me: I'd rather wear a T-shirt to hide my perceived imperfections than open myself to judgment. I thought I'd escaped the clutches of fatphobia, but in an intimate moment, doubt roared in my ears.

I'm far from the only one who has succumbed to doubt in a vulnerable moment—even Oprah Winfrey has felt the pressure to change herself. In 1998, then thirteen years into her reign as America's favorite—and richest—television host,

Winfrey was asked to appear on the October cover of *Vogue*. During a 2009 interview with Morley Safer on *60 Minutes*, *Vogue*'s famed editor in chief, Anna Wintour, said she had "gently suggested" to Winfrey that she lose twenty pounds before appearing on the cover. "I went to Chicago to visit Oprah, and I suggested that it might be an idea that she lose a little bit of weight before she appeared in the magazine," said Wintour. "She was a trooper. She totally welcomed the idea. She went on a very stringent diet. It was one of our most successful covers ever." While Winfrey has never confirmed that Wintour asked her to lose weight, she did say on her show that "giving [herself] over to the *Vogue* experience" meant that Wintour told her, "'You had to release all control to me.' . . . Now I worked my behind off, literally," Winfrey said. "People say, 'Where's the beef?' It was on my behind. Over the summer I was working out, working out, to get in shape. And it paid off. I was transformed into *Vogue*'s October cover girl."

The demand that plus-size women lose weight to appear on magazine covers, or risk being excluded altogether, isn't new, and it has persisted up to and through Ashley Graham appearing on *Vogue*'s March 2017 cover with seven other models. Graham is the first plus-size model to grace *Vogue*'s cover, but it is notable that she didn't appear alone, which seems to be a trend. In October 2015, *Elle* celebrated its twenty-fifth anniversary with four different versions of the cover, featuring Megan Fox, Amanda Seyfried, Gabourey Sidibe, and Lauren Conrad. While Conrad's, Fox's, and Seyfried's covers showcased their full bodies, Sidibe's cover zoomed in closely on her face, so she could be seen only from the neck upward. Not only was *Elle* accused of purposely hiding Sidibe's body, but they also allegedly lightened her skin. "By cropping Sidibe's cover photo so close, *Elle* may have been trying to hide her full-figured body—its own travesty—but they only made her seem bigger," Julianne Hing wrote in a September 2010 story for *Colorlines*. "Sidibe doesn't get the standard female cover photo treatment: three-quarters of the woman's body centered with strong margins of white space on either side of the woman. She gets a uniquely awkward cropped shot."

Similarly, when Melissa McCarthy shared *Elle* cover honors with Penélope Cruz, Reese Witherspoon, and Shailene Woodley, she was also posed differently. McCarthy's calf and a triangle-shaped area near her neck were the only visible parts of her body, while Witherspoon was decked out in a curve-hugging Versace dress and Woodley wore a skimpy black swimsuit that highlighted her toned arms and trim figure. Of course, *Elle* denied the obvious implications of, once again, hiding a fat woman's body on their cover. "On all of our shoots, our stylists work with the stars to choose pieces they feel good in, and this is no different: Melissa loved this look, and is gorgeous on our cover," a spokesperson for *Elle* said in a statement at the time. "We are thrilled to honor her as one of our Women in Hollywood this year." McCarthy also defended her cover: "What I found so bizarre is I picked the coat. I grabbed the coat. I covered up. I had a great black dress on but I thought, it comes out in November," she told *E! News* in October 2013. "I was so sick of summer. I live in Southern California. I was like, 'Give me a big coat to wear. Give the girl some cashmere!'"

While we must respect McCarthy's insistence that she not only chose her outfit but was comfortable with being photographed differently than her fellow cover stars, the decision to give her a different, more modest wardrobe is indicative of a bigger problem with the ways in which magazines—and the media at large—understand and treat plus-size bodies. Our bodies are too imperfect to be revered, plastered on magazine covers in two-piece swimsuits. Hiding until we've reached the appropriate body mass index (BMI) score is the only option presented. Or, as Whitney Way Thore, the star of TLC's *My Big Fat Fabulous Life*, told *Splinter* in 2016, plus-size women are purposefully mistreated on magazine covers because so many publishers believe that showing a fat woman's full body will decrease newsstand sales.

"I do definitely think there is a tendency to minimize plus size women so that we are flattered, and slimming," Thore said. "It makes them more digestible and easier to handle so that people still want to buy the magazine. But what's the point of putting her on your cover if you don't want to celebrate her body? It piggybacks on what we've always heard: 'You're beautiful, but you ought to cover your stomach' . . . it's not necessarily blatant, but as soon as you point it out, it really does send a message." That message is clear: as fat women, we should cover our bodies, lest we corrupt innocent people with our stretch marks and rolls.

For a long time, I assumed I hadn't internalized the message that plus-size bodies should be camouflaged in oversize clothing or hidden altogether. After all, I had no qualms about being naked, even when it meant changing in front of other people. But in one of those moments that are marketed to women as "special"—losing your virginity—everything I'd seen on those magazine covers and in pop culture fell on me. Keith and I still had sex nearly every day, before classes, between classes, and after classes, over the weekends, during and after studying—nights on nights on nights to make up for the time I'd spent having relationships over computer screens. And every time we lay down, my shirt stayed on and the lights stayed off. He was patient, as he should've been, as I slowly unfurled like a flower, opening emotionally as I opened my legs the first, second, third, twenty-fifth, and fiftieth time. Even as I let him into the most delicate, intimate, and vulnerable parts of myself, I kept my shirt on and the lights stayed off—until he began slowly pushing me to fully unrobe and bare myself to him.

He was measured, gentle, and understanding, waiting for cues that signaled to him that I was ready to literally and figuratively get naked. Unveiling the body I worked so hard to hide was a process; it started with him coming into the bathroom to brush his teeth and wash his face while I showered. He pretended as if he weren't watching me bathe behind his glass shower door, engaging me in ordinary conversation about my plans for the day and what I wanted to eat for dinner. Then, as I turned the shower off, he'd leave the bathroom—pushing, but never too much. Up until that point, I'd undressed, showered, moisturized, and changed into pajamas or an outfit in the bathroom, but his gesture—small, but thoughtful—encouraged me to stop hiding.

It took a few weeks or maybe longer before I began coming

out of the bathroom in a towel, lotioning in the room instead of the bathroom, and allowing him to see my stretch-marked, flabby, fat body. Though I was now comfortable enough to dress in his presence, the T-shirt still lingered during sex. If I was wearing a dress, I'd roll it up just enough to give him access to the goods, but never enough for his body to lean on mine without a barrier.

After we'd been having sex for more than six months, he carefully broached the issue. "Why don't you take your shirt off?" he asked quietly between kisses, his warm hand resting on my breast. I quickly shook my head no and tried to continue kissing him, but, instead, he sat up, looking down at me with a puzzled look on his face. "Please take your shirt off," he said again, with just a little more forcefulness in his voice. I covered my face with my hands, afraid to face him and admit my fear of losing his attention, his desire, his attraction to my imperfect body. I again shook my head no and then tried to sit up and reach for my pants; the charade was up, and so was our relationship in my estimation.

Fatphobia can make us think irrationally about ourselves, our bodies, and the moments in which both are being scrutinized, even when that scrutiny is positive or rooted in love. Instead of letting my irrational thoughts drive a wedge between us, Keith gently pushed me back down, removed my hands from my face, and then looked into my eyes as he lifted my shirt. He pulled the T-shirt off, laid his body on top of mine, pushed my knees apart, and slid inside me in a few swift motions. By the time it fully registered that I was having sex with my T-shirt off, I was already cumming and no longer concerned about Keith still wanting me. He removed my shirt the next time and the time after that and the time after that until it became just a part of the way we had sex. I no longer felt the need to hide, and he no longer would allow me to. Yet, coming undone in the privacy of his room or mine didn't initially translate to other areas of my life.

To most of us summer means warmer weather and outdoor activities: visiting beaches, spending full days at amusement parks, frolicking at pools, or eating picnic lunches at parks buzzing with people and vendors. For me, summer has also meant wearing less clothing. It's one thing to wear crop tops and mini-shorts; those revealing items can be paired with looser garments to create an illusion. Two-piece swimsuits offer no shelter; sun rays reveal the flab, cellulite, and stretch marks, which embarrassed me on beaches packed with "ideal" bodies. And like many fat people, I worried that venturing into public spaces, like pools, would result in my being taunted or made to feel uncomfortable.

That fear is warranted, as evidenced by preschool teacher Jacqueline Adan. After losing 350 pounds, Adan decided to wear a bathing suit for the first time while vacationing in Mexico. As soon as she removed her cover-up on the beach, a couple began laughing at her because of the loose skin that comes with significant weight loss. "I did not know if I was going to cry or put my cover-up back on and not go swimming. For a moment I froze," Adan told *People* in September 2017. "I could not believe how someone could judge me and make fun of me. They had no idea who I was, what I had been through or what I was currently going through."

While Adan ultimately decided that their opinions didn't matter, she encountered similar criticism in 2018 after undergoing skin removal surgeries, as she discussed on her Instagram: "It happened this year while I was on vacation again," she wrote. "Yes, I was walking on the beach and again I was pointed at and laughed and made fun of. As I was getting laughed at and pointed at something came over me. I was not embarrassed, I did not feel like I had to justify myself and I did not freeze or want to cry. I actually felt free! This time . . . I just did not care! I tried to go over every change inside my head as to why this year I just did not care. Then it finally hit me. I do not depend on the approval of others, and I do not care what others may or may not think about my body."

I LEARNED VERY EARLY ON THAT COVERING UP MY BODY WAS THE ONLY way to be comfortable at pools and on beaches. I'd always wear a one-piece under a swim dress or a T-shirt that I'd peel off only when I was steps away from entering the water. It was a frustrating and delicate dance; I had to remove the shirt at just the right time so people didn't have time to gawk. In college, as my friends and I traveled to Myrtle Beach and Miami, I grew tired of hiding at the beach and trying to stuff my wet body back into a T-shirt after getting out of the water. But I didn't have the confidence to make a different decision until fashion blogger and designer Gabi Gregg, a.k.a. Gabi Fresh, made it her mission to take back the beach. In April 2012, Gregg set the internet ablaze when she ventured to Las

Vegas with her boyfriend wearing a black-and-white-striped two-piece bikini from Simply Be.

At the time, a few brands had started designing chic plus-size swimsuits, including ModCloth, Monif C., Evans, and ASOS, but the fatkini was still a nascent idea. Gregg helped catapult plus-size swimsuits into the stratosphere. She looked so comfortable and chic, posing as if she were wearing a flowing tunic or another garment I've used to camouflage my midsection. I was astounded by her confidence, and so were many other women on the internet. Two months later, Gregg partnered with the now-defunct blog xoJane to curate a gallery of thirty-one other fat women wearing "skimpy swimwear," and within a year, she'd started her own affordable and cute capsule collection with Swimsuits for All to encourage other women of size to push aside their insecurities, step outside of their comfort zones, and showcase their beach bodies.

"As always, I truly encourage you guys to get to the beach (or a pool) this summer—don't let body shame keep you from having a good time!" Gregg wrote in a blog post when she first went viral. "I can't tell you how freeing it is to just have fun without worrying about what other people think." Though I admired Gregg's confidence from afar—and I wished I could bottle it, drink the elixir, and inhabit the same spaces that she did—I didn't have the no-fucks-given approach that she so effortlessly embodied. Instead, I spent one dreadful and lonely summer in Minneapolis without friends, without a romantic partner, and without the confidence to go to one of the city's numerous, famed lakes in just a swimsuit.

As I remained resistant, a burgeoning movement of plus-size bloggers, including Gregg, began pushing for fat women to embrace their bodies and parade their curves on beaches. They were working to reach women like me, women who couldn't move past the fear. It seemed that their goal was to create a community of bloggers, designers, and ordinary women so that fat women could be affirmed by those who were already on the journey to body acceptance and wanted to usher others into the fold. "I think what changed my perspective and what helps so many women get to a place of self-acceptance is seeing other people that look like them," Gregg told *Teen Vogue* in 2013. "This could mean going to Tumblr and looking through the body-positive and plus-size fashion tags so you don't feel so isolated. Girls who grow up as the 'fat girl' in their class can feel really alone, so seeing other women who look like you do and are happy and stylish can really change the way you see yourself."

In 2014, as Gregg's #fatkini boomed into a movement, with more than twenty thousand plus-size women tagging their photos with the hashtag on Instagram, Shameika, my childhood best friend, encouraged me to get on the train. We were shopping for her twenty-fifth-birthday cruise, which would travel from Miami to Key West to Cozumel, Mexico. Every time I screenshotted a photo of a bland one-piece swimsuit and sent it to Shameika, she'd counter it by sending me a photo of a bolder two-piece swimsuit in a flamboyantly bright color. Each time I tried to purchase another cover-up or swim dress that would hide my body, she'd either ignore me or send me a semblance of "Bitch, where are you going with that on?" Heeding her advice and taking the risk of buying a swimsuit that looked cute on plus-size models who weigh less than 200 pounds was one of the most difficult things I've ever done, but is there any better time to take a risk than when you're on a cruise ship with thousands of people whom you'll likely never see again?

I was nervous when I stepped onto the ship's Lido deck in a neon-green two-piece with no cover-up, but with Shameika's encouragement, I pushed negative thoughts aside to allow positive reinforcement to emerge. Instead of wondering if people were taking sneak pictures of me as I tanned beside the pool, I closed my eyes and reassured myself, "You are beautiful. You are loved. You are confident. You will never see these strangers again." When I worried about wearing a bright pink halter swimsuit to snorkel on a beach in Cozumel, I focused in on the experience itself instead of the potential criticism and shaming that might be aimed at me from strangers. By the end of the trip, I'd never been surer about any fashion decision I'd made. Being comfortable in a fatkini reminded me that women of size have every right to flaunt their bodies. We have no reason to be ashamed, even when the entire world tells us we're "glorifying obesity" by displaying them as thinner women do.

Even plus-size fashion lines emphasize that particular message. Season after season we see collections that include few flourishes, outside of statement sleeves, ruffled hems, and the occasional crop top that shows just enough belly to be considered risqué. This only emphasizes the backlash levied against the fat women who choose to defy the cultural rules that are never spoken but are wholly understood. When women of size wear fatkinis or showcase their bodies on red carpets or take off their T-shirts during sex, they are brazenly casting away a fat-shaming culture. And now, there's an entire movement that revolves around plus-size women challenging the idea of a "beach body," telling the beach that it will get whatever body fat women give it, and scores of swimsuit collections that prioritize the comfort of fat women above all else. These swimsuit designers don't attempt to pull off fashion tricks that camouflage the "unflattering" parts of fat women's bodies. There are mesh and side cutouts and long V-necks and all the options that used to be exclusively available to thin women.

When Gregg released her fourth swimsuit collection with Swimsuits for All in 2016, she told *Splinter* that her goal was to continue encouraging plus-size women to disrupt places that were previously unavailable to us unless we bent to the fatphobic will of shamers and critics. "I'm always really conscious of wanting to push the boundaries and wear things that people say I shouldn't wear," she said. "And that's in all clothing, but in swimwear I think it's really important to not feel ashamed of your body and be able to show the same amount of skin as everyone else."

Being unashamed is an ongoing lesson I'm still learning and may always be learning. More than ten years after I had sex for the first time, I am still learning to be comfortable with being nude in the presence of a romantic partner. I stopped having sex with my shirt on long ago because it's both uncomfortable and makes for an awkward conversation I'd rather avoid, but in the moments before a partner's skin melds into mine, there's always a second when I pause and worry about whether this will be the time when I'm finally rejected. It's never happened, and one day, I hope to move past this irrational fear. But in the meantime, I'm stepping into the nude—and leaving that T-shirt on the side of the bed. Finally.

THE SKINNY BOYFRIEND TROPE

Laying my eyes on Elijah for the first time in 2012 is a memory that's forever seared into the folds of my mind: I stood there frozen as he slowly walked across my college's gravel-strewn parking lot, dressed head to toe in black with the brim of a black baseball hat covering his brown eyes. We'd connected through a dating site, so I knew that he was a plus-size man with a gorgeous, joy-inducing smile, neat locs that hung just beneath his shoulders, a well-oiled beard, and an incessant need to compliment me every time we exchanged messages on the dating app, texted each other, or spoke on the phone. But envisioning the size of someone's body based on perfectly angled photos designed to attract a partner on an online dating site is vastly different than confronting their body offline. It's different than measuring them up, deciding if there's an attraction—and grappling with the fatphobia that often dictates how we understand, relate to, and foster romantic relationships with fat people.

As Elijah walked closer and closer, and the start of our first date loomed, several paranoid scenarios flitted rapidly through my mind: Would one of my classmates or friends see us together and taunt me for dating a fat man? Would someone secretly snap photos of us and then spread them on social media as a way of poking fun at us—a fat couple? Preemptive panic and fear set in, and I found myself unable to walk toward Elijah's gray sedan, as I'd originally planned to when I'd left my dorm room. I waited, seemingly forever, for him to reach me, and when he did, all the paranoia dissipated into the warm night. I looked him up and down, noticing his beard and the largeness of his belly and waist. At that moment, a sense of calm set in. I felt safe with him. Though we'd met online, I felt confident he wouldn't butcher me—turn me into one of those victims whose stories run on Investigation Discovery marathons.

"I'm Elijah. Are you Evette?" he asked as he reached out his hand to me. I nodded quickly, looking up at his looming six-foot-two frame, still unable to find the words, but able to place my hand in his. "Can I hug you?" he asked. I hesitantly nodded again, and then looked around, making sure nobody was lurking in the shadows, before I stepped into his arms and embraced him. I knew I wouldn't be able to fit my arms around his waist, so I opted for his neck, and hugged him— first with hesitancy, and then with eagerness. Nobody was watching, so I hugged him and then hugged him again. As long as we were in this secret bubble, away from prying eyes, I was invested in getting to know him and hopefully, if he was as affectionate as he seemed when we texted and talked on the phone, committing to a monogamous relationship. I was preparing to graduate from college in four weeks, so I thought Elijah could fulfill the age-old dream of meeting the love of your life during those four years and living happily ever after. At that point, it didn't really matter who the man was; I just wanted to live the dream.

After we'd hugged, and I'd inhaled his scent—a mix of weed and a clean-smelling cologne—he enclosed my hand in his and walked me across the parking lot and to his car. That was the closest Elijah ever got to step foot on my college campus; I relegated our relationship to the parking lot, where there was less of a chance we would be spotted together and all my fears about being seen together would come true. We'd met on one of those dating sites that many college students frequented in the 2010s—it could've been Plenty of Fish or Tagged—and I was creating profiles faster than I could respond to messages. At the time, I was still in the middle of ending a relationship, but I couldn't quite get to the end. Although Miguel lived less than two hours away from my Greensboro, North Carolina, campus, he could never fit me into his "busy" schedule. We'd plan a date. The day would come, and he'd cancel or just simply not show up, but this was always followed with an excuse such as getting locked up after a drunken night that included pissing on the sides of buildings. After more than six months of being disregarded, and in some respects ghosted, I decided to move on and find someone who wanted to be in a relationship that included face-to-face dates.

Elijah was one of the first men with whom I matched when I began using dating sites again, and I was immediately intrigued by him. He, unlike so many other men I've encountered, had a profile that was completely filled out. He had multiple photos that displayed his locs and his smile and a bio that explained why he had moved from California to North Carolina, and he spelled out his intentions clearly without any pretenses. Elijah was open to casual relationships but really wanted a girlfriend who would eventually become his wife. I swiped right because we had similar interests, an equal investment in being in a committed relationship, and a lot of ambition that could be encouraged through a friendship and, eventually, a romantic partnership. After we sent a few messages through the app and exchanged phone numbers, Elijah took the initiative to ask me on a date.

This was the opposite of what I'd experienced in my previous relationship; his urgency impressed twenty-one-year-old me. But when he stepped out of his car in that parking lot, I felt repulsed, and then overcome with shame for having such a viscerally fatphobic reaction to his plus-size body—since I was also fat. Dating sites are often cesspools for plus-size people, especially for fat women. We're often fetishized or pursued by "feeders," people who overfeed fat people and enable us to overfeed ourselves to make us undesirable to others. (Yes, this a real thing!) Or else we're being fat-shamed—whether it's through those "no fats" or "prefer women who exercise" (as if plus-size people don't use their bodies for anything other than eating) that litter profile bios or being deemed liars and catfishes because we choose to show our faces instead of our entire bodies.

In 2014, the YouTube channel Simple Pickup launched a

"social experiment" with a woman who wears a seventy-pound fat suit on Tinder dates. Each time she meets a potential suitor, she is faced with a cringeworthy reaction that demonstrates the extent to which dating apps can ostracize fat people. "You look kind of more voluptuous," one of her dates says before asking, "Are you pregnant? I really don't appreciate people lying to me." He immediately leaves. Another man tells her, "I'm a little upset. I wasted gas and my time to come over here, and I can't do this." The video also cites a study that found that men most fear being matched on a dating site with a woman who's fat. Online dating is so bad for the fat community, in fact, that in 2016 the website WooPlus was launched specifically to cater to fat people who are seeking online romance. Michelle Li, one of the cofounders of WooPlus, was inspired to create the site after watching Simple Pickup's video. "There was a lot of mocking, and basically making her feel lesser of a person because of her size," Li told *The Daily Dot* after WooPlus launched. "We thought, 'that happens a lot.' We thought why don't we come up with a dating app to see just what kind of result we'd get out of it. We're just trying to provide a comfortable environment for women who happen to be a little larger. It's for women of all different bodies and sizes. But we wanted to show that you don't need to have a low BMI [body mass index] to be attractive."

Men have crudely messaged me on dating sites about "wanting to eat my fat ass," have called me "thickness" and "fat queen" in introductory messages, and have even told me that I should lose weight before trying to date. As much as that rejection stings, however, I have also perpetuated sizeism against potential partners on dating apps. I've swiped left on men simply because they're larger than me or because they, like me, have more than one chin. Dating doesn't exist in a vacuum of choices that are somehow separate from the ways in which we've been socialized to think about desirability and attraction. As queer, nonbinary, Black, fat, femme cultural producer, data futurist, and multidisciplinary visual artist Hunter Ashleigh often says, dating is political.

Whom we find attractive and whom we choose to date is a reflection of our indoctrination into a culture that creates hierarchies of desire around race, gender, religion, and size. But knowing how attraction is colored by oppression doesn't excuse the shallowness that often guides my dating decisions, especially on apps. It's difficult to unlearn fatphobia in a culture that upholds only the kinds of men who dominate #BigGuyTwitter (husky and muscular with no flab) and #BodyPositive Instagram accounts (pinching one inch of fat and calling it oppression) as attractive. All the messaging I internalized about fatness shone through my relationship with Elijah. While I've written screeds about it being fatphobic and discriminatory to hide relationships with plus-size women, I participated in the very culture I rebuked.

Our first three dates were so delightful that I became convinced I was meant to be in a relationship with Elijah. Yet, I couldn't express this to my friends, my classmates, or my relatives because they'd already started making insensitive jokes about our courtship. I downplayed our burgeoning relationship and laughed along with them as they pointed out that we were a mismatched couple because we were both fat.

My friends bombarded me with uncomfortable, penetrating questions: How would we have sex? Did I worry about eating in public with him since we were both fat? Would we have fat babies? With shame and embarrassment blossoming in my belly, I swallowed the discomfort and humiliation, answering their questions as if we were all in on the same joke. The only party who wasn't privy to the framing of our relationship as a humorous bit was Elijah; I was into him but not enough to shut down the sizeism being perpetuated against us.

Broaching the topic, calling them out, and holding them accountable for the fatphobia that had crushed Elijah and me our entire lives was too difficult, so I chose to confront the trauma in private. I'd internalized so much fatphobic rhetoric that dating Elijah seemed like a sin. At the outset of our relationship, he was incredibly doting and devoted: he would prepare my favorite meal at the time—fried pork chops, macaroni and cheese, and broccoli—and bring it to me in that parking lot I never let him move beyond. He randomly brought me flowers, surprised me with thoughtful gifts, and even helped me decide which graduate school to attend. But on the night he asked me to pack a bag because he'd planned something special, my competing desires—wanting to be intimate with someone I cared about and wanting to avoid being the butt of the joke—clashed. Elijah rented us a beautiful hotel room for the night, complete with a Jacuzzi tub, a table with a spread of some of my favorite foods and desserts, and rose petals that led from the door, through the small living room, and into the bedroom. He'd put so much effort into making the first night we were supposed to have sex special. But even as we ate the delicious food he'd prepared and he massaged my neck and my shoulders, gently tilting my head back so that he could softly plant his lips on mine, those haunting questions from my friends and family crushed my chest like a large boulder that I didn't have the strength—or willpower—to remove.

After dinner, separate showers, and a movie that neither of us paid much attention to, Elijah made his move; there was kissing, the fondling of my breasts, the removal of my night-gown, the rubbing of my feet, the sucking on my clitoris, the lingering heat between us, and me pushing him away as soon as he tried to climb on top of me. I blamed the rejection on him not having a condom, but I knew it wasn't just the lack of protection that made me turn onto my side, wrap the blankets tightly around myself, and whisper that I just couldn't have sex with him. "Are you serious?" he asked as he stood up and pulled on his shirt and boxers. "I am only with you. You're my girlfriend, and you know I'm not fucking anybody else. Why do we need to use a condom?" I closed my eyes tightly, and quietly reiterated that I couldn't have sex with him without protection. He sighed before grabbing his car keys and leaving the hotel room, making sure to slam the door behind him.

My mind raced as I waited for him to return to the room. What if he didn't come back? What if he went to get condoms? What if I ran out of excuses? I decided that falling asleep was the best option because maybe, just maybe, my mind would be clearer in the morning. Eventually, Elijah came back to the room with condoms, but as he tried to nudge me awake, I purposely sank deeper into sleep because I wanted to avoid what would inevitably come next. I ignored all his efforts to wake me up, until he finally sighed with frustration, turned onto his side, and began watching another movie until he, too, fell asleep. When we woke up in the morning, shame seeped through my skin and into the blankets. How could I be so repulsed by someone whom I cared about? How could I deny sex to someone who treated me with so much care and tenderness? No matter how much I tried to reason with myself, I couldn't move past the size of his body. I couldn't have sex with him. Though he seemed to be the partner whom I'd been searching for, I couldn't envision a future with Elijah.

There were few things I wanted more at the time than a successful and healthy relationship with Elijah. We'd had several really intimate conversations about our goals and our dreams, and our visions seemed to align. We'd casually discussed our future: He'd move to Illinois with me once I started graduate school, and when I was finished, we'd move to New York to pursue my dream of being a magazine editor. In the meantime, he'd be pursuing a career as a hip-hop artist and producer and working security for local nightclubs. We talked about getting married, having children, and buying a condo. Love drew us together, but my inability to confront my friends' and family's fatphobia drove a wedge between us.

I was a willing participant in perpetuating the fatphobic idea that Elijah and I weren't meant to be together because of the size of our bodies. When I told my friends about turning down Elijah's advances in the hotel room, I framed it as a joke. "I thought he was going to crush me. There was no way I was letting him climb on top of me," I said with glee, taking joy in their laughter. "It has been fun, but we have no future together," I continued. "I can't fuck a fat dude." He was the punch line in my cruel jokes, and I knew that he always would be. My relationship with Elijah was a reminder of the extent to which social factors like outside judgments, societal expectations, and—of course—what we see on-screen influence how we treat one another.

As a 1994 study, "Growing Up with Television: The Cultivation Perspective," found, the more exposure we have to media images, particularly on television, the more we attach real-world expectations to what we're seeing. Being raised in a fatphobic culture where I never saw fat couples on-screen and rarely saw them in real life influenced my inability to see past Elijah's size and engage in a healthy partnership with him. Plus-size couples don't exist—at least not in popular culture. While the Centers for Disease Control and Prevention estimate that 36.5 percent of adults in the United States are "obese," a pesky term that doctors use to tell you to lose weight every time you come into their office, TV shows and movies are primarily populated by thin actors who don't mirror the actual population. Multiple studies have quantified the impact of never seeing plus-size characters on-screen. For example, a 2003 study published in the *American Journal of Public Health* found that plus-size women characters were less likely to date, have a romantic partner, or display romantic affection on-screen. Given how infrequently fat women, particularly fat women of color, appear on TV, it's no surprise, then, that fat couples are a rarity on television.

Often, we're treated to mixed-size couples—a plus-size male star partnered with a smoldering woman who is seemingly ill fitted for him. Take Doug Heffernan, played by the affable Kevin James, in CBS's long-running *The King of Queens*, for instance. Doug's a delivery driver for the fictional International Parcel Service who lives in a modest house in Queens, New York. As writer Matt Feeney points out in a 2005 article for *Slate*, he's often unable to complete simple tasks because he's so obsessed with remedying personal grievances at the expense of those around him.

Doug is clumsy, untrustworthy, and often cruel, but somehow, he marries Carrie (Leah Remini), a drop-dead-gorgeous legal secretary with a temper and a mean streak that scares their neighbors. Since Doug is fat—and is often taunted about his size by his wife and his friends—he becomes obsessed with making sure Carrie retains her looks and figure.

His ability to have a "perfect" wife seemingly makes up for all the tormenting he endures, and he projects those insecurities onto Carrie. In season 1, he persuades Carrie to go on a diet after he sees an old picture of her now deceased mother. He hates when she wears her hair in a bun, so he uses petty means to persuade her to wear her hair in curls. And after a waitress flirts with him at their favorite restaurant, Doug starts frequenting the establishment, just to make Carrie jealous. The humor of their marriage is that Doug is plus-size but obsessed with how his wife looks, while Carrie is gorgeous and combats his attempts to control her by being cruel.

Unfortunately, their dysfunctional on-screen relationship is the rule rather than the exception. Mixed-size couples dominate television. On ABC's *According to Jim*, Cheryl (Courtney Thorne-Smith) and Jim (Jim Belushi) are the primary couple. There's also Peter (Seth MacFarlane) and Lois Griffin (Alex Borstein) on *Family Guy*. Philip (James Avery) and Vivian Banks (Daphne Reid) on *The Fresh Prince of Bel-Air*. Gloria Delgado-Pritchett (Sofía Vergara) and Jay Pritchett (Ed O'Neill) on *Modern Family*. Hell, even Miles (Jack Black) and Iris (Kate Winslet) in one of my favorite films, *The Holiday*, are representative of this trend.

When we do see plus-size couples on-screen, the caveat is that their relationship must revolve around their bodies—more specifically, on their mutual quest to make their bodies smaller. What we're given are cautionary tales camouflaged as love stories. In *Mike & Molly*, when Chicago police officer Mike Biggs (Billy Gardell) meets fourth-grade teacher Molly Flynn (Melissa McCarthy) in an Overeaters Anonymous meeting, they bond over their inability to lose weight. In fact, as Molly walks into her first meeting, Mike is at a podium oversharing about his desire to gain control over what he eats so he doesn't become an old, lonely fat person with "six or seven cats who happened to wander into my gravitational field." While *Mike & Molly* is one of the rare sitcoms that stars a plus-size couple who have a functional, healthy relationship, the show is full of fat jokes and focuses far too much on their struggle to become thin.

In one grueling scene, for instance, Molly is exercising on a stair climber as her mom, Joyce Flynn (Swoosie Kurtz), with whom she shares a home, watches. "You're always going to be the big-boned girl," she tells Molly in an effort to comfort her. "You've just got to accept that you've got your daddy's genes. I mean, if you had a turkey leg in one hand and one hand down your pants, I'd swear he was risen from the dead." The show attempts to poke fun at the experience of being fat, which is depicted as being more about navigating jokes than facing the fatphobia that forces Mike and Molly into Overeaters Anonymous to begin with. But this simply accentuates what fat people know to be true: if we're going to find love, especially with another fat person, then our relationship has to revolve around an elusive quest to become smaller.

Molly speaks to this when she tells Mike that she's not interested in losing weight per se but wants to control what she eats (as if one doesn't feed into the other). "I know I'm never going to be a size 2. I just want to learn to control my eating," Molly says early in the first season. "I'd love to be able to walk into a nightclub without having every queen in the room leaping on me like I'm a gay-pride float."

Another plus-size couple on television are Kate Pearson (Chrissy Metz) and Toby Damon (Chris Sullivan) on NBC's smash drama *This Is Us*. Just like Mike and Molly, Kate and Toby meet in a fictional weight-loss support group, because apparently that's the only place where two fat single people cross paths. The beginning of their relationship, too, completely revolves around their desire to lose weight. When Kate realizes that Toby wants to pursue a romantic relationship with her, she tells him she can't date another fat person because it will distract her from losing weight. Eventually, he wins her over, but just as soon as their relationship really gets started, Kate decides to end it because Toby leaves the weight-loss support group. While their relationship seemed beneficial when they were encouraging each other to follow diets and exercise, it's as if their shared commitment to being smaller was the only thing holding them together.

Toby tells Kate he will eat the same foods she does when they're out on dates, but she decides to dump him anyway. "I think we should take a break because I've been dreading this date," she tells him as he eats a plate of spaghetti while she eats a salad. "The other night, I stuffed my face full of powdered doughnuts, and I cried myself to sleep. I told you that when I met you, that I couldn't fall for a fat person." In season 4, Kate and Toby have to persuade a gynecologist to allow Kate to undergo in-vitro fertilization because the doctor considers Kate too high-risk, due to her size, to get pregnant. Of course, their son, Jack, is born premature and nearly dies—all because his parents are fat. When we see plus-size couples, like Mike and Molly and Kate and Toby, it's clear that every aspect of their relationship deals in some way with their weight. They are forced to perform contrition for seeking happiness while not fitting into a thin ideal. Despite all the trauma written into their stories, love sustains their union, which is something fat women need to see on-screen. We need to see a fat woman being supported and nurtured by a partner who could care less about her ability to sustain weight loss.

Still, there's a larger repulsion attached to seeing fat people on-screen. In 2016, Maura Kelly wrote a much maligned and now retracted blog post for *Marie Claire* about *Mike & Molly*. "No one who is as fat as Mike and Molly can be healthy," Kelly wrote. "I think I'd be grossed out if I had to watch two characters with rolls and rolls of fat kissing each other . . . because I'd be grossed out if I had to watch them doing anything. To be brutally honest, even in real life, I find it aesthetically displeasing to watch a very, very fat person simply walk across a room—just like I'd find it distressing if I saw a very drunk person stumbling across a bar or a heroine [*sic*] addict slumping in a chair." When you are as aware, as most fat people are, and as I definitely was when I was dating Elijah, that you are under constant surveillance—when you eat, get on a plane, go to the doctor, or even try to fit into movie theater seats—the goal becomes shaking off the target that's on your back. Choosing to partner with someone who's also facing scrutiny merely because of their size takes courage—a courage I didn't have.

"There's nothing worse than seeing two fat people together" has rolled off the tongue of many people I've encountered, including my own family and friends. Can two plus-size people have a successful relationship? Yes, but it requires practicing the difficult but necessary art of tuning out insensitive commentary about your relationship. On our first date, Elijah took me to one of those chain restaurants that sell good drinks and appetizers for reasonable prices. Immediately, a waiter tried to sit us at a table at the center of the restaurant that had high barstool chairs. In a calm but firm tone, Elijah told our confused waiter we couldn't sit there and pointed out where he'd prefer we sit—a spacious booth in the back corner where we wouldn't be on display for other restaurant-goers. Our waiter sheepishly obliged.

Navigating those very simple elements of relationships, like where to sit when we went out for dinner, was made difficult by a world that's not built to accommodate larger bodies.

A fear of becoming a public spectacle—a fat woman dating a fat man—made me push Elijah away for nearly two years after our hotel room debacle. He was supposed to meet my parents and attend my graduation ceremony, but I made excuses for why he couldn't. I ignored his phone calls and text messages, and then I pretended I didn't remember that we'd made such arrangements. When I left North Carolina two weeks later to spend the summer in Minneapolis and then head to graduate school in Illinois, it became very easy to leave Elijah—and the memories we'd created—behind. He never brought up that night in the hotel room and I never did either. I buried it instead, content to use the narrative of long-distance relationships being too difficult to maintain as an excuse for why our relationship so quickly eroded.

When we separated, I could sense the relief from my family and my friends; they no longer had to worry about me having "fat babies" or being stared at as we walked down the street. In their eyes, ending my relationship with Elijah was a return to normalcy—for everyone but me. Six months after we broke up, I sent him a message on Facebook as I walked a treadmill in my personal trainer's private gym. I couldn't shake Elijah or the affectionate moments we had shared or that quiet inner voice telling me he hadn't deserved to be shunned in that hotel room. He quickly responded to my message, and we fell right back into a routine, as if no time had passed at all. We abandoned our old plans of moving in together and decided to continue a long-distance relationship until I finished grad school. While outwardly I expressed disappointment about us not being able to move in together, internally I breathed a sigh of relief. I didn't have to worry about integrating him into my small college town or about being ostracized. And when I visited him in North Carolina in April 2013, one year after I'd graduated, we had sex—a lot of it. We locked ourselves in a hotel room for nearly a week, only coming out to get food, and in the privacy of our own space, I was able to be myself and really push aside all the negative thoughts about our bodies.

Good sex wasn't enough to sustain our long-distance relationship, which we decided to end for good in June 2014. Nor did it put an end to my battle with the internalized fatphobia that dictated so much of our relationship. Even now, years after we've separated and I've learned about fat acceptance and the politics of desirability, I still find myself swiping left on the majority of plus-size men. It's unclear if I'll ever be able to partner with a fat man, or even if I deserve to, but I'm at least on an ongoing journey to unlearn fatphobia—and Elijah is still the conductor who first put me on the train to doing so.

YES, MOTHERS CAN BE MONSTROUS

Imagine being whistled at, as if grown men can no longer form words that ordinarily come to them with little effort. Imagine the constantly honking horns, not directed at traffic, but at you, a woman just trying to get home safely. Compliments yelled out of windows that quickly devolve into insults when the "romantic" gesture is rejected. Being followed and walking more quickly to escape the harasser, who's matching your every stride. Smiling politely and reminding the person glaring at you that you have a boyfriend, and no, you're not interested in dumping him. Thanking him for the compliments, moving out of his hand's reach, fake-smiling as you walk away, and hoping he doesn't follow you home. If you're a woman, chances are you've experienced at least one of these scenarios, which exist on a spectrum that ranges from innocent flirting to death.

It started innocently enough for me. After working eight hours and commuting an hour from Manhattan to Brooklyn,

I ascended twenty-two concrete steps, stopping at the top of the subway station to catch my breath before crossing the street to catch the bus that would drop me near the brownstone apartment I'd recently moved into. What I was wearing doesn't matter because it never matters. Whether I'm wearing a short skirt or a long one, cloth shorts or jeans, a long-sleeved shirt or a halter top, there's always a man who believes my personal space can be invaded, as if the boundaries encompass enough room for two.

I spotted him out of the corner of my eye, just in my periphery, leaning out of the tinted driver's window of his gleaming black car. I intentionally stared straight ahead, careful not to acknowledge his increasingly frantic attempts to get my attention, lest he think the interest was mutual. I continued nodding my head to the music pumping loudly through my earbuds. I ignored his incessant and belligerent yelling. I ignored his frantically waving arm. I don't know why I thought that he'd get the hint; street harassers never do. Then, he decided that ignoring him wasn't an option for me. He pulled away, tires screeching, and swerved his car into a parking lot directly across the street from the bus stop. For a moment, I thought he'd abandoned his quest to get my attention and was going to the bodega instead. But when he exited his vehicle, he headed straight toward me. My heart did a rapid one-two dance, loud enough that it was pounding in my ears as if it were preparing to leap out of my chest. I continued to ignore him, but he didn't take too kindly to that. He walked right up to me and planted his feet directly in front of me. I had no choice but to pretend as if I were just seeing him and his antics for the first time.

"I've been trying to get your attention for like ten minutes," he said animatedly, his hands moving to keep up with his words. "You're really cute. Can I take you out sometime?"

"Oh really? I didn't see you. And thank you, but no. I have a boyfriend," I replied with a small, nervous smile. I looked to the left, hoping the bus would pull up at that exact moment.

"Your man isn't okay with you having male friends?" he asked as he licked his dry lips. We went back and forth like this for seemingly forever—with me politely rejecting his advances and him aggressively dancing around my disinterest. He sidestepped every mention of my partner, tried to persuade me to let him give me a ride home, and even tried to force me to give him my phone number. When the bus finally arrived, I breathed a sigh of relief as I quickly shuffled up the stairs and into the back, where he could no longer stare at me. Even talk of a partner can't protect you from a man intent on using public space to remind you that he has the power to own and control you.

There was also the man with long, pressed black hair who followed me in his rusting Bronco truck as I walked down Brooklyn's cobblestoned Eastern Parkway. Whenever he thought he'd lost me, he'd speed up, at one point almost hitting an elderly white man pushing a silver cart across a side street. When he was finally able to pull over and block me in, he told me I was beautiful, and if I ever dumped my boyfriend, I should give him a call. That day, I was wearing a knee-length green bubble coat, the kind with insulated lining that protected my body from whipping winds. Though I repeatedly expressed my disinterest, and the cars behind him honked as the traffic light turned green, he refused to drive off until I accepted a business card that listed his name and phone number. As soon as the card graced my fingertips, he pulled off, and once he was out of eyesight, I crumpled the card and dropped it in the nearest trash can.

Inconsiderately delaying traffic and offering tired lines about beauty is supposed to be flattering, especially for women of size, who, the logic goes, have been so deprived of romantic attention that even hinting at attraction or desirability should be enough to open our hearts and our legs. In this world where beauty is built on hierarchies, fat women are pinned at the lower end of the totem pole, where few suitors are willing to venture, so even if the flirting is entwined with harassment, it's better than being invisible. Right?

For many years, I considered street harassment a normal, everyday part of every woman's life, as routine as crowded trains running late in New York or contracting a cold every winter. Being hassled by strangers was just the price paid for being a woman in a patriarchal world. For me, this kind of unwelcome attention began when I was a teenager.

When I was sixteen, my mom, a family friend, and I were going into a beauty supply store to pick up black gel, bobby pins, and whatever else Black girls need to secure their fake ponytails. I can't remember exactly what I was wearing, but I do remember an older Black man leering at me, wrinkles embedded in his forehead, from the moment I stepped out of the car. "Hey girl, you're the most beautiful woman I've ever seen!" he said incredulously while my mother and my mother's friend giggled uncomfortably. "It should be a crime for anyone to look that good!" My cheeks flushed red with embarrassment, and my eyes found the concrete as we walked into the store.

Less than five minutes later, he approached me as I walked down an aisle with a pack of bobby pins in my hand. He tried to reach for my hand, muttering about how beautiful I was, but my family friend stepped between us to remind him that I was a child—and I solely belonged to myself. He leered again, looking me up and down, before exiting the store. I shook it off, smiled, laughed along with other people in the store who thought this encounter was normal, necessary, and a boost to my sixteen-year-old ego. My mother chuckled as we exited the store, delight twinkling in her eyes, because her baby, her fat teenager who spent so much time conversing with strangers on the internet, was finally considered attractive enough to be hit on by an older man who smelled of alcohol. But I just felt afraid and confused as I tried to understand what could possibly be entertaining about the humiliation that had seeped into my skin, wrapped itself around my bones, and unraveled me from the inside out.

"Oh girl," she said dismissively as I opened the back door of our burgundy Intrepid and climbed into the backseat. "There's nothing wrong with being appreciated by a man." I blinked but didn't respond, hoping that staying silent would help us both move past the moment.

It was the first time street harassment left me feeling shaken and unsafe and the first time my mother revealed her hand. In her mind, there's real value in being desired, even when the encounter is unwanted or violates personal boundaries.

All attention is good attention, and women should always engage with men who shower them with it. My mother is one of my closest confidantes, the person who I turn to the most, and our relationship is also complicated. She, like me, has been fat the majority of her life—since, as she likes to say, she got her tubes tied in her late twenties. Unlike me, however, she has focused the majority of that time on slimming her body. She's seemingly tried every fad diet and infomercial exercise machine in existence, hoping that this time she'll lose the weight and keep it off. And once she's reached a modicum of success, whatever benchmark she's set for herself, she'll turn to me and casually try to persuade me to join her on the diet or use the machine or purchase a gym membership. "Don't you see how well this is going for me?" she'll ask, as she sneaks cookies and chips throughout the day. "Don't you want to do it, too?"

It isn't malicious; in fact, there's nothing else my mother has ever known. In a world that indoctrinates fat women into a dieting death cult and then socializes them to see their bodies as a work in progress that just requires a little more discipline to be fixed, her relationship to food, to dieting, to exercise is par for the course. It's so par for the course that when my cardiologist instituted a low-sodium diet and liquid restriction, she barely blinked before she switched out all the seasonings in our cabinet and began mixing up low-sodium dishes and experimenting with our favorite recipes to make them as low sodium as possible.

Heart failure has ebbs and flows. I am lucky to be on a stable medication regimen that has been showing improvement through quarterly echocardiograms, or heart ultrasounds, which can reveal how well blood is being pumped from the heart into the lungs and throughout the body. That doesn't mean it's without cost. One of the concerns, always, is that when the left ventricle fails—as mine did—the right side of the heart will soon enlarge and fail. The heart is almost like an ACL: once you tear one, you can bid adieu to the other one as well. However, from early on in my cardiology treatment, it seemed as if my right ventricle was holding steady. As I stuck to that strict medication regimen and liquid and sodium restrictions, my cardiologist marveled at how much better my heart was performing.

My ejection fraction—the amount of blood pumped through each contraction—was 16 percent when my heart failure was caught. Quarter by quarter, we watched it turn itself around, to 25 percent, to 45 percent, and eventually to 55 percent—or low-normal—a heart capable of pumping enough blood to sustain my body without causing fluid to pool in my lungs and in my ankles. Over the course of a year, I'd gone from being unable to walk up steps without losing my breath to being able to walk for thirty to forty-five minutes without breaking a sweat. The improvement felt like a triumph—until it didn't. Though my left ventricle was improving and my cardiologist had even begun sending me to see a nurse practitioner instead of her, I wasn't feeling as good as I thought I should feel. My lungs still burned when I overexerted myself. I still never felt well rested. Even when I took big breaths, I didn't feel as if I was pulling enough air into my lungs. I started to panic, so I pushed my cardiologist to give me another echocardiogram— and that's when it all came crashing down.

My right ventricle had begun enlarging, which meant it would eventually fail. It had gone unnoticed because my cardiologist hadn't been monitoring the full heart. Instead, she'd been ordering echocardiograms only for the left side, so my right side was allowed to enlarge without any medical intervention. Once I'd found a new cardiologist (because of course I did), a series of diagnostic tests made the underlying cause of the right enlargement clear: pulmonary hypertension, an aggressive and progressive lung condition characterized by high blood pressure in the lungs that thins the veins and makes it difficult for oxygen and blood to pump through. Eventually, it causes the right side of the heart to fail. It was happening to me and I could feel it. On my heart failure journey, I have often felt as if I'm suspended in the ether and watching someone else—an avatar, if you will—living out a chronically ill existence. When I learned that I was still in the danger zone despite all of the progress I'd made, it finally hit me: This is really happening to me. I have heart failure. I have pulmonary hypertension. I can't self-optimize my way out of being ill.

When you're first diagnosed with a chronic illness, one of the first things doctors tell you is not to google your symptoms or your conditions. I'd always heeded that advice—until I was diagnosed with pulmonary hypertension. It's a rare condition, striking less than two hundred thousand people a year, so I wanted to soak up as much information as I could as quickly as I could. It was all doom and gloom, as to be expected: Poor prognosis, worsening conditions, essentially drowning in your own lungs. Doom. Doom. Doom. But underneath the fearmongering, there was one kernel of truth:

pulmonary hypertension is a contraindicated condition for pregnancy.

There is a 50/50 chance that someone with pulmonary hypertension who becomes pregnant will die, their child will die, or both parties will die. Every pulmonologist and cardiologist recommends that people with pulmonary hypertension avoid becoming pregnant at all costs; some doctors even require pulmonary hypertension patients to take birth control and a monthly pregnancy test before agreeing to treat them. Of all the things I've learned about the conditions I've been stricken with—all the risks, all the damage they cause to the body— learning that I'd be unable to carry my own children hit me harder than I ever could've expected. Motherhood wasn't a societal state I'd ever pined for. I'd always thought about it, of course, in the usual way that middle-class women are socialized to think about it: Figure out a career path. Find a partner. Get married. Purchase a home. Have a couple of children. Learn how to juggle it all. But I'd never thought further than that about being a parent. If it happened, it happened, and if it didn't, it didn't. Living in this peaceful limbo was satiating for a while—until I realized I might not have the opportunity to mother at all.

MOTHERHOOD HAS NEVER FELT AS URGENT FOR ME AS POP CULTURE AND our larger social pressures suggests it should. I'd never dreamed about having an extravagant wedding, birthing 2.5 children, or purchasing a home with the standard white picket fence, and since the urge hadn't overtaken me, I'd imagined it never would. But when I learned that there was a possibility I'd never be able to birth my own children, all the considerations I'd brushed aside came flooding back. While I may not want a husband—and the baggage that often accompanies marriages, even healthy ones—I do desire a child (or two), children who will be raised in an untraditional family, one comprising primarily a single mother, grandparents, and a fuller village of aunts, uncles, and siblings, some biological and some chosen. My sudden interest in being a mom caught everyone I love by surprise because I just hadn't foreseen that being a part of my life. But when I made the decision, it felt like the most natural thing in the world. I could be a mom and a damn good one, even without a husband.

When President Joe Biden selected Kamala Harris to be his running mate, midway through one of the most unconventional campaign seasons on record, political pundits marveled about all the ceilings she'd be breaking, all the firsts she'd achieve as America's first South Asian and Black woman vice president. Regardless of the controversy that surrounded her candidacy—being California's attorney general and enforcing the state's laws comes with a whole lot of harm—one of the unique aspects of her appointment to one of the nation's highest political offices is how she formed her family. In a country that purports to care about "family values," and that has often considered two-parent homes with two cis-het parents to be not only the standard but the *only*, Harris's family looks much more modern American than her predecessors' families have.

Harris married her husband, Doug Emhoff, when she was

in her forties, after her best friend set them up on a blind date. He'd already been married and had two children from that union, so Harris and Emhoff opted not to have children. She also opted to keep her maiden name. She knows who her husband is; her name doesn't have to reflect that. Whenever Harris, who is the child of Jamaican and Indian immigrants, takes the stage, she's often flanked by her sister, her nieces, and their families—a fully blended family that isn't defined just by those who live in her home. They're an interfaith and interracial family, led in part by Kamala, a.k.a. "Momala," a stepmom, a sister, and a great-auntie. Their family, in many respects, represents what's possible in a time when so many women are heavily invested in our educations and our careers.

When my children are born via surrogate, an already untraditional choice, they will be entering a world that's making strides toward accepting different family formations but still asks for a male partner's name on the forms I've had to complete as I've undergone egg freezing. They'll be asked, repeatedly, about their father and forced to navigate the scrutiny that comes with not having the traditional two-parent home. Though I can fill their lives with activities, with resources, with a village that loves them, am I setting them up for pain by choosing to approach life in this way? Am I being selfish? Have I internalized the idea that only "nuclear" families are worthy of being considered, respected, and honored? Am I setting my Black babies up for failure? If racist history is any indication, then yes.

In 1965, Assistant Secretary of Labor Daniel Patrick Moynihan released *The Negro Family: The Case for National*

Action, better known as the Moynihan Report, which was designed to persuade President Lyndon B. Johnson not to sign vital and necessary civil-rights legislation. Within that report, Moynihan stated that "the deterioration of the Negro family" is the "fundamental source of the weakness of the Negro community at the present time." It wasn't just that Black Americans were being subjected to abject racial terror as well as discrimination in every facet of their lives; it was a moral failure that their families didn't look like the white families that graced prime-time network television. To Moynihan, this was the reason the family structure was breaking down—and Black women were to blame.

In Moynihan's view, working Black women "undermine[d] the position of the father and deprive[d] the children of the kind of attention, particularly in school matters, which is now a standard feature of middle-class upbringing." Everything about the condition of the Black family fell on Black women's doorsteps. Though Black women have made incredible gains in terms of education and the workforce—we're now the most educated group in the United States—we've also seen our romantic prospects dwindle for a number of reasons. I've watched Black men online and offline pick Black women apart: They're not marrying us because we're too loud, too aggressive, too disagreeable, too independent. Who would willingly subject themselves to that kind of vitriol? We'd rather be alone.

Many Black women have made the decision to forgo romantic partnership and pursue motherhood on our own terms—stereotypes be damned. I'm as confident as I have ever been that the world will accommodate my chosen singleness. After all, I'm a six-figure earner with the means and the resources to hire help. I might not be able to attend every one of my children's plays, but I can guarantee that someone in my family will be present. I'm surrounded by other high-earning people who can fill in the gaps, monetarily and otherwise, whenever needed. But that doesn't mean my children won't be subjected to the idea that I'm selfish for creating an untraditional life for them. I don't have children yet, but I've talked about them as if they already exist, in the pages of *Essence* magazine, on social media, and to the people in my life whom I love the most. I've committed to a lifetime of raising those children, no matter what it takes, because I so deeply believe in the idea that we should all have the right to have the family we desire. I can decide not to subject myself to emotional terror from men, not to put myself through disappointment. But can I tell my child that having a father doesn't matter because I've deemed it so? That answer remains unclear.

EVEN BEFORE I'D BEEN DIAGNOSED WITH PULMONARY HYPERTENSION, aspects of pregnancy and motherhood—conceiving, gestating, and laboring in the traditional way—had been slowly chipped away at. It started with fibroids in 2014. Three in four Black people with uteruses will develop fibroids in their lifetime, but usually they are asymptomatic. But I would bleed and bleed and bleed for days and sometimes weeks at a time. I was a full-time educator then, teaching eighth graders who didn't miss a beat—ever—and we were embarking on a field trip to visit a local college. I was wearing light pink jeans and I'd wrapped a jean jacket around my waist because I knew I'd bleed heavily through the entire trip. But even though I was changing my menstrual pad every hour on the hour, that didn't stop blood from pooling in the seat of my pants. As we approached the bus on our way back to the school, one of my students quietly said, "Miss, you have blood in your pants." The embarrassment of a student telling me that truth—as well as the shame I'd been swallowing about fibroids signaling something defective about my uterus—persuaded me to seek a medical intervention.

Within a month, I was having a robotic myomectomy to remove a fibroid while also keeping my uterus intact, because I wanted to retain the possibility of carrying my own children in the future. That first myomectomy stole the potential of having a vaginal birth from me; a surgeon told me that I'd always need to have a scheduled C-section, or I'd risk rupturing my uterus and killing my child. My second myomectomy, in 2019, stole everything else—small piece by small piece. The only thing that stopped the bleeding was a hormone called megestrol acetate—the same hormone that attacked my heart and caused it to fail. As part of my treatment, my cardiologist and pulmonologist prescribed me a bronchodilator, a medication designed to keep the blood vessels in the lungs loose and flexible, allowing blood to flow more easily. The prescription came with a clear instruction: pregnancy was strictly prohibited because the medicine causes severe birth defects in children.

This presented me with a dilemma. Pulmonary hypertension is a lifelong condition, so I'd likely be taking this medication for the remainder of my life. Therefore, I would never be able to be pregnant. The same hormone that treated my fibroids had caused my hypertension. And now, it had taken away any possibility of birthing my own children.

In that moment, everything clicked for me. I may not be able to carry my own children, but I want to be a mother. And if I wanted to retain that option while getting the best possible treatment for my condition, I'd need to freeze my eggs for an eventual surrogate before beginning the medication. I made the decision overnight. Within twenty-four hours, I'd scheduled a consultation with a fertility specialist. Within one week, I'd signed all the proper forms to begin the mature oocyte cryopreservation—or egg-freezing—process. Three weeks later, my mother—the same woman I'd come to understand as ensnared in the same fatphobic systems I was—was injecting hormones in my abdomen every night to increase the quantity and size of my eggs. Thirteen days and countless needle pokes later, I had seventeen eggs removed from my body in an outpatient retrieval surgery, thirteen of which were deemed viable and are now hanging out in a freezer, waiting to be fertilized with sperm and implanted in a surrogate.

If my gynecologist had immediately referred me for a myomectomy rather than putting me on megestrol acetate, I wouldn't have had to shell out nearly $15,000 to freeze my eggs and the close to $100,000 it will cost for surrogacy. I'd be able to carry my children in my own uterus, though pregnant Black people face their own sets of dire circumstances. You need not look further than maternal mortality rates in the United States to know that's true.

Serena Williams is one of the most powerful women in the world. Depending on who's asked, she's the greatest athlete of all time. But she's undoubtedly one of the most successful tennis players of all time, with twenty-three Grand Slams to her name. She's also been on the front lines of a movement to raise awareness about America's maternal mortality crisis, which disproportionately impacts Black people—in fact, Black people are four to five times more likely to die during childbirth than white people. There are a variety of reasons for this: systemic lack of access to quality health care, rampant poverty, and medical neglect from doctors carrying around anti-Black bias. Williams is one of the wealthiest Black people in the United States and has access to some of the best-quality health care in the world, but giving birth still nearly killed her. When she birthed her daughter in 2018, she knew she was experiencing blood clots from a previous condition, and she had to keep calling out for help from doctors tasked with keeping her alive in order to receive the surgery she needed. That has everything to do with her Blackness.

There's no doubt that I will be a fat mother, just as my mother was a fat mother. I imagine I will birth children who will grow into bodies similar to mine, meaning I will likely raise a fat child or two. That's what scares me most. In our cultural imagination, fat mothers are absent, neglectful, or abusive, with little space for three-dimensional characters who are simply a positive force in their children's lives. Lee Daniels's Academy Award–winning film *Precious*, based on Sapphire's 1996 novel *Push*, offers the most extreme example of the supposed harm that fat mothers cause to their children. Mary, played by actor and comedian Mo'Nique—who also won an Oscar for the portrayal—treats her daughter, Precious, as if she's merely a burden who can't become an adult fast enough. Precious is an illiterate and HIV-positive single teen mother living with her mother in public housing in New York. Her father is also the father of her two children, and though he's no longer in the picture, his specter still hovers in Precious's life—after all, she's raising his children-grandchildren.

In addition to the trauma Precious is navigating as an incest and rape survivor, she's forced to face Mary's relentless emotional, mental, and physical abuse alone. Mary wants to condemn Precious to her same fate, so she doesn't encourage her daughter to educate herself and she's resistant to her daughter receiving any outside help from teachers or social workers. Mary's miserable—for reasons that the film never attempts to explore or explain through a backstory—and she cares more about retaining welfare benefits than attempting to help her daughter secure some semblance of a more prosperous life. It's impossible to watch *Precious* without feeling overwhelming anger toward Mary for being so neglectful and abusive, which is seemingly the point. No reason is given for why this fat, dark-skinned mother treats her own daughter so poorly; it's not because she's also endured trauma in her childhood, or because she feels trapped in social conditions that she's made to feel responsible for, or because she feels hopeless.

Mary's simply abusive seemingly because it's inherent to her nature, an idea that's inextricable from the literal body she inhabits. For decades, we have seen similar depictions of fat mothers in pop culture, usually without any deeper interrogation. In the 1993 film *What's Eating Gilbert Grape*, the title character (played by Johnny Depp) is forced to be the primary caretaker for his disabled brother, Arnie (Leonardo DiCaprio), because his mother, Bonnie (Darlene Cates), is literally eating herself to death. She's grown so large, in fact, that she's losing her mobility and she will soon be completely bedridden if she doesn't lose weight.

Thanks to Bonnie choosing the pleasure of food over rearing her children, Gilbert is shouldering all the responsibilities that should fall on her: cooking, cleaning, repairing the house, making sure Arnie—who has a fear of water—bathes, and keeping their small town's gossip mill as much at bay as possible. It's Bonnie's neglect that becomes the catalyst for the film, as Gilbert struggles to balance all of the expectations placed on him with the life he's trying to create for himself as a young man. He's freed from the struggle only when Bonnie dies alone in her bed. Rather than giving Bonnie a proper burial, Gilbert burns down the house with her in it because he calculates that that is better than having to answer questions about what killed his mother.

Our cultural performance of motherhood builds itself upon the idea that mothers—and those who fulfill that role—must be selfless, must overexert themselves, must be self-sacrificial, and must be all-knowing, all-seeing, and all-nurturing. Fat mothers who buck these conventions, especially if they are Black, are marked in pop culture and our broader society as deviant and irredeemable. That is likely the reason Bonnie's punished, dies without cause, and then is not properly buried.

I want to raise children who know I love them unconditionally, no matter what size their bodies are. I want to have children who are confident in their skin, full of joy, and concerned more about their grades than about adhering to a diet culture that idolizes thinness. What I fear, however, is falling into the same traps my own mother did—trying to model confidence in a society that teaches children early on that it's okay to pick on fat children. I fear my children having the same experiences I did as a child, feeling targeted and surveilled at school and in doctors' offices and feeling as if their home is the only safe place to land. I fear my children being ashamed of me because I can't do everything their classmates' parents can do or because they weren't conceived and brought into the world the way the majority of babies are.

I will be a chronically ill mother, who will often have to prioritize my own health needs above the immediate needs of my children. There are going to be times when it will be impossible for me to cook dinner for my children, chase them around our neighborhood park, and get up early enough to greet them before they leave for school. The literal limitations of my body won't even allow me to birth them, an idea that often leaves me feeling unspun, as if surrogacy is my punishment for falling ill and remaining ill. I'll never be able to take maternity photos. I'll never be able to attend prenatal birthing classes. I'll never feel my child kick in utero. I'll never be able to go into labor and experience the joyous moment when my child exits my womb to come into the world. I'll never have a birth story to share during holiday dinners.

But what I do have to offer—a village full of people who

will love my children, a heart full of love and support to give, a willingness to be humbled by motherhood and guided by the very beings I bring into the world—may outweigh everything else. It's the "may" that keeps me up at night. I've lost so much already, before motherhood has even begun. And becoming a mother will continue to cost me financially and emotionally. I imagine it will be difficult for me to watch someone else carry the child I wanted to carry. But I am more than willing to pay these emotional penalties in order to be a fat mother. In fact, it's the biggest sacrifice I'll ever be willing to make.

THE IMPERMANENT PROTOTYPE

Men tell me I'm the prototype. They whisper it in my ear as we fall asleep, breathe it into my mouth as their hard flesh sinks into the softness of mine. It's a sweet nothing designed to gently ease my guard down, remind me that this time—the hundredth time—will be different. It never is. Men tell me how beautiful I am, how ambition radiates off my skin like dewy sweat on a hot day in the middle of summer. They tell me how much they admire me, that I'm brilliant, that I'm kind and loving and thoughtful—the woman they'd love to bring home to their family because they know their parents will approve. They tell me, in so many words, that I'm the prototype—the woman they never expected to meet but suddenly have within their grasp. Nobody tells you the prototype is the version before the final. The prototype is simply the blueprint—the floor model—not the one brought home and nurtured, cared for. The prototype is experimented on: let's see how much this one can endure, so we can make sure the actual woman hurts less than she does.

I can't recall the exact moment I realized I was the prototype, always disposable and never permanent. It might have been in 2010, when my second sex partner ditched me for a night out with his friends—and then ignored my text messages for a week. He'd told me I was the woman he'd dreamed of, but he couldn't commit, because he wasn't ready. We could go on dates and cuddle at night. He'd even introduce me to his friends, but me being his girlfriend was beyond his scope of comfort. We had sex twice before I stopped answering the phone when he called. Maybe it was in 2011, when my third sex partner claimed we were building a future—and then refused to be seen in public together. When I finally blocked his phone number—after he'd tried to humiliate me on Facebook by sharing a secret I'd told him in confidence—I was so depressed I couldn't get out of bed. My friends had to force me to eat, to shower, and to go to class. Maybe it was in 2012, when my fourth sex partner kissed me so sweetly that I just knew—in my bones—that our relationship would eventually blossom into something beyond sex. Two sex sessions later, my bones said, "Nope, not him," and I cut him off, too. He lived thirty minutes away, so he'd usually pick me up at night and drop me off in the morning, but when he said he'd prefer for me to have a friend pick me up, I knew he was only in it for the sex. He didn't care about anything besides what was between my legs. Maybe it was in 2019, when a married man didn't tell me he was married and then lied about the state of his relationship in order to keep me dangling at the end of a twisted rope.

Sex is the great equalizer. In my experience, being intimate with someone creates the greatest clarity. When you first meet a person and realize there's a legitimate connection, the lust builds and "forever" seems as if it's a possibility. You hear what they say through a distorted filter, one driven by a craven need to feel their body on yours, in yours. When sex finally happens—no matter how long it takes—you suddenly see things more clearly than ever. You are good enough to only be laid down with. After all, they say, aren't you beautiful and ambitious and brilliant? Everything said in the in-between—commitment isn't on their radar, they'd prefer to be nonmonogamous, running through women is stitched into the very fabric of their masculinity—suddenly snaps into focus. The connection wanes, as spotty as AOL dial-up in the late '90s. I am the prototype—meant to be used as a guiding light for the person coming to them after me.

The women they'd ultimately choose didn't resemble me at all. Many of them were thin, light-skinned, or ethnically ambiguous, with long hair swinging down their backs. I never considered that these men just weren't ready for a relationship and I was asking them to be, perhaps unfairly. I just assumed there was something defective about me, primarily because I'm fat. The end always came soon after the beginning because my expectations came into focus—and men don't meet expectations they're not interested in meeting. I wasn't perfect in these scenarios, especially when I was in my twenties:

I expected these men to know what my standards were, even when I didn't outwardly communicate them. I expected them to be a boyfriend without ever asking them to be.

I wish there were a way in which being continually rejected in this way—or ending things before I can be disappointed—didn't chip away at my self-esteem and self-assuredness. After all, I am a Black feminist existing, learning, and loving in the tradition of Patricia Hill Collins, bell hooks, Joan Morgan, and Audre Lorde. Love, as hooks defines it in *All About Love: New Visions*, is an "act of will, both an intention and action." It's the "will to extend one's self for the purpose of nurturing one's own or another's spiritual growth." Love doesn't just exist within the perimeters of romantic relationships, and it isn't rooted in possessing another person. It's an experience. I inherently know this. It's the tradition I've been raised in, the one that has shaped my worldview. I know exactly who I am, and I know all I've unlearned to reach a point wherein I value myself outside of how much I'm desired by men. And yet, in a culture that rewards cis-het women for being chosen by a man—a single man who conveys his love through an engagement and then through wedding vows—remaining unchosen takes a toll. It becomes easy to internalize rejection. Why am I not good enough? Why have I been unable to conform enough? Those questions have been specters, long hovering above my life and driving me to tears—very quiet ones that I shed at night, alone, after realizing, again, that it's going to end because it's destined to.

In 2009, there was a lot of hubbub about single Black women being unable to find spouses. The often touted statistics that 42 percent of Black women in the United States have never been married and that there are 1.8 million more Black women in America than Black men left an open question that everyone from Steve Harvey to *Nightline* and *The Washington Post* tried to answer: What's wrong with Black women? Are our dating standards too high? Are Black men dating interracially because we're not submissive enough? Should we date interracially because Black men are? Setting aside the fact there are queer Black folks who aren't interested in cisgender, heterosexual Black men, this hyperfocus on single Black women pathologizes our experience. It frames singleness as a curse that we must exorcise to attain any semblance of happiness. It also ignores the fact that some Black women just aren't interested in being married at all. What becomes of those of us who want to opt out altogether?

When the disappointment subsides, the tears dry up, and the hurt retreats, I always come to the same finish line: Being single doesn't mean I'm lacking in love, but it does mean that I can protect myself from the unique hurt that accompanies the ending of a romantic relationship. I'd rather be single. Singledom is not something cisgender, heterosexual women are socialized to desire. What matters most is being partnered with an elusive soulmate who just inherently understands you. When you're unable or unwilling to find that special one, you're written off—just another single Black woman incapable of conforming to the norms our society has prescribed us.

Although I've been dating since I began college in 2010, most of the men I've encountered are simply blips in my memory. If I concentrate hard enough, I can recall some of their facial features, but if I blink, they disappear as if they never existed. In this flip-book of unimpressive, forgettable men, there's one who stands out: I can still see the contours of his chocolate face as clearly today as I did when we first connected in 2011. He's the "one who got away," who escaped my clutches before we could establish a foundation for a relationship and build upon it. Gary and I met in college, and I thought he was perfect. We fit hand in glove, as if we'd known each other our whole lives. He was handsome. He was funny and he made me laugh—a rarity. He was ambitious. He loved God and he loved his family. He wanted to be a pharmacist—or a professional pill pusher, as he called it—and he wanted to be a husband. He wanted children. He said he wanted me. We lived in two different cities within the same state, so we'd spend our nights on the phone, both clacking away on the computer and cracking jokes in between finishing our homework. It was fun and easy—until it wasn't.

It started innocently enough: Six or seven months after we'd begun communicating, I received a notification that he'd uploaded a new photo on Facebook. He was seated at a table in a restaurant and flashing his beautiful smile, but his pose, the easiness in which his face relaxed, drew a question for me: Who took the picture? When I asked him, he fumbled and bumbled before sheepishly admitting that he'd been out with another woman—and he was considering committing to her. I forced back the water that gathered in the corners of my eyes, wished him well, and retreated.

As so many men do, he popped back up eight months later to tell me he'd made a mistake. I was beautiful, brilliant, and ambitious, and understood him better than any other woman he'd ever met. He missed laughing with me. I let him back in—that time and the next time and the time after that—because I thought that all the tears and the hurt were merely a sacrifice in this ongoing quest to be chosen. When he married the last woman he'd dumped me for, on my birthday, my heart broke for the final time, and I developed a steely resolve in my belly. Rather than opening myself up to the possibility of being hurt or being heartbroken, I'd prioritize everything but a partner. I'd be the best writer I could be, the best editor I could be, the best daughter and sister I could be, the best aunt I could be, the best friend I could be. I'd fill my world up so much that there would be no room for someone to weasel their way in, gas me up with promises they didn't intend to keep, and abandon me when they realized I was simply the prototype. In fact, I'd created such a complete life for so long that even considering making space for a partner—having to merge two lives, having to compromise, having to create a new reality together—seemed both impossible and immensely undesirable. What could a man provide to my life beyond stress?

There are penalties for bucking convention—loneliness at times, isolation on occasion, invasive questions—but it's also a freeing experience. Rather than waiting for a man to choose me, I chose myself. I chose to build the kind of life I desire: abundant, prosperous, successful, and full, without once considering if this life will be amenable to someone trying to enter it. There's no need for a soulmate or even a partner, because I have the resources to create the life I desire without the need for a second income or an additional person's input. If my children need support, they have a village of people who will show up. If I'm feeling depressed, there's a legion of girlfriends on whom I can rely to pull me through. If I want to travel, I can do so without negotiating with a partner or working around his schedule. If I'm craving sex, I can either provide it to myself or call up someone I can put out after. For as much as singledom is discouraged—and as much freedom as it has provided me—I return to the same question over and over again: Why aren't women encouraged to be free?

I NEVER INTENDED TO BE SINGLE. IT WASN'T IN MY PLAN AND I NEVER ENVIsioned a life without a partner. When I first met Marcus, I was an isolated twenty-three-year-old grad student navigating a major depressive episode without the language to describe the complex feeling of being upended and turned upside down. Nearly every night, I'd climb into my full-size bed, turn on my television, and let the screen watch me as I debated whether I should get re-dressed and drive to the local McDonald's. I'd usually cave around 1 a.m., throwing on leggings and a sweatshirt to slink out of the apartment I shared with a single roommate and driving slowly down the dark road to avoid deer, the illuminated, elevated golden arches beckoning me in.

When I wasn't soaking in my apartment's larger-than-normal tub for hours at a time, devouring true crime shows, and fighting with myself about what I "should" eat, I was literally putting on a happy face. By that final semester, I had only two classes and I was a teaching assistant for a single class, so most of my time was my own. While this light schedule was supposed to give me the time to research and write my thesis, the lack of planned interaction only sank me further into my depression. Going to class and being a TA were the only times I was social, so I prepared for it each time. I'd inherited the warped idea that I should never look as bad as I feel, so I'd take a long shower, pluck a matching bra-and-panty set out of my dresser, iron my clothes, put on a little eyeliner and bronzer, slide on a cute pair of heels, and throw on a few sprays of my most expensive perfume. I knew I was depressed; nobody else needed to know.

I'd strut into every class with my head held high and a smile plastered on my face. Greeting people felt easy in those academic spaces because we had a shared topic to converse about. "What do you think about last week's reading?" was often my opening line. It was the only real time I felt "normal," though I knew I might go home after class and bawl until the skin under my eyelids felt raw and sensitive to the touch. When I marched into a giant lecture hall on the first day of the final semester, I immediately spotted Marcus. He was so tall and so lanky that he couldn't fit his entire body into a single seat. Instead, he hung his legs over the seat in front of him. During those awkward first-day introductions that make us all cringe, he stood up and shared that getting tattooed was his hobby—and it showed. He was drowning in ink, with both of his arms nearly covered and tattoos peeking over the collar of his white T-shirt. I was intrigued, but I knew he was off-limits. After all, he was a student in a class I was TAing.

I knew nothing could come from this sudden infatuation,

but Marcus didn't seem interested in those boundaries. Whenever the class met, he'd greet the professor teaching the class and then immediately greet me, smiling and raising his eyebrows. Whatever outfit I'd put together—first for myself and then, over time, for his approval—he would visibly take note of and appreciate. Nearly every class, as the teacher delivered her lecture, he'd ask me to sit beside him. He'd pat the seat beside him and look directly in my eyes, almost attempting to hypnotize me into saying yes. It took six weeks of invitations—and six weeks of his fellow students laughing as I rejected him—before I agreed. I was so lonely. Anything would do. When I took some of the students to see a Cornel West lecture, he tagged along, holding my hand to help me walk up the stairs to our assigned balcony seating and repeatedly asking me if I was okay as I shifted my fat body to try to get more comfortable in the metal seats. We left the lecture together, looking up at each other and exchanging smiles as we walked toward the lot where we'd both parked our cars.

As I climbed into my car, I knew the boundaries were out the window. I was going to fuck this man. In the small life I'd been confined to, being intimate with him was all I had to look forward to. I knew I'd be violating an unspoken rule about sleeping with a student, but I somehow justified it: "We're both adults. I'm not directly teaching him. Nobody will ever have to know." A few days later, I went through my customary dressing ritual, spending extra time in the shower buzzing down my pubic hair, and then headed to the library. I didn't know how I knew Marcus would meet me there, but in the pit of my gut, I knew I'd cross paths with him that day.

He slunk into the library a little more than an hour after I'd gotten there and made a beeline for the table where I was sitting. We watched a basketball game together on my laptop, exchanged numbers, and, eventually, left the library together for my apartment. That was the last time we were ever truly separated. I knew it was complicated. He was still in a relationship, and I was ending an unsatisfying long-distance relationship with Elijah, the fat man I once rejected, but Marcus and I were like magnets to each other. One week later, he was helping me drive to Detroit to grieve with my family after a death of a cousin. After graduation, I made the quick decision to stay in our small college town for the summer so I could finish writing my thesis and see if something, anything, could blossom between us.

We quickly fell into a comfortable pattern after he and his then girlfriend broke up and I ended things with Elijah: He'd go to work in the morning, come to my apartment for lunch and sex, finish his work shift, and then spend the night. When he wasn't working, we'd spend our afternoons fishing at the school's lake, visiting wineries, and hiking—all the things he enjoyed doing. Hiking made me feel tired and achy, but I never complained, because we were doing it together. When he'd come through during his lunch break, I'd ask him what he wanted for dinner and I'd either pick it up or make it, even if that wasn't what I wanted to eat. When he wanted to hang out with his friends, I'd drive to pick him up when he was too drunk to drive himself to my home. When he was passed out on my room floor and his phone was vibrating, I'd open it to see him promising other women the world. When his friends were hungry, I'd feed them, too. While I didn't feel so depressed that it was difficult to get out of bed, it was only because my boyfriend had become my whole world. I didn't know where I ended and where he began; we were melded, one. Midway through the summer, Marcus and I drove to Denver together to meet my parents and see if we could make the city a home.

That had never been my plan before. After I finished grad school, I thought, I would move back to New York City and make my way through the media industry, but that wasn't what Marcus wanted for us, so I tucked that idea away, deep in the recesses of my mind and my soul. He fell in love with Denver during our trip, so I pretended as if I had as well. And at the end of the summer, when both of our leases were up, we packed up our cars and headed out west to begin a new life together. We moved in with my parents for three months before leasing an apartment, purchasing furniture, combining our bills, merging our lives. We adopted two dogs. He went to work at our neighborhood airport while I began teaching eighth grade. We were happy. We were always happy. Our social media grids were full of pictures of us adventuring and smiling together. We radiated joy. But while he was content with the small life we'd begun building for ourselves, I'd slowly become miserable. We were always together. I always prioritized his needs above my own. I paid most of our bills, and when I couldn't afford our expenses, I went to increasingly extreme lengths to maintain our lifestyle. I pawned the iPad my parents had gifted me for graduation. I'd buy it back and then pawn it again. I even pawned my car to pay our rent.

Marcus was such a good partner and good man that I just wanted to appease him in every way I could, no matter what it personally cost me. He was always so encouraging. Hiking with him had helped me lose a noticeable amount of weight. His mother and his brother treated me as if I were one of their own. I'd never been loved in public before or truly loved by a partner in any way, so I was committed to keeping his attention. When I wasn't orgasming anymore because he refused to invest in foreplay, I faked it. When he started interrogating me every single day about what I'd eaten while he was at work, I adhered to his demand. I never wanted to upset him and, therefore, we never argued in the two years we lived together. I smiled, even when I was crying inside. But as is often the case, depression rolled up on me without warning. It started with sneaking into the school bathroom every day after I finished teaching to quietly cry, my body shaking with the sobs I didn't dare give voice to for fear of being found. I would then cry as I drove toward home, stopping at a fast-food restaurant on my way, eating in the car, and dumping my trash before I walked through our apartment's front door. The last thing I wanted was for my weight-obsessed boyfriend to chide me for not eating a salad for dinner. No matter how deep my sadness ran, I'd paint a smile on my face when Marcus came through the door. I was down, but he didn't need to be.

I'd been repressing so much of the stress, pressure, and sadness I felt that the dam eventually broke. The tears I compartmentalized when I was alone were suddenly leaking all over, and I couldn't hide them. I'd bawl in the morning as I got dressed and cry myself to sleep, turning my back to Marcus so he wouldn't see my upset and attempt to fix it with a "you'll be just fine" platitude. Within a month, I couldn't get out of bed. I felt stuck—in my teaching career, in my relationship, in my body, in every facet of my life. I inherently knew that, but I didn't have the language for it. Instead, my tears were symbolic of the depths of my discontent. During those difficult times, Marcus became the partner I'd always hoped I'd have. He stepped in. He helped me apply for a leave of absence from teaching. He began cooking every night when he realized I couldn't do it. He'd run me a bath and then coax me to get into it. He helped me find a counselor. He held me every night, refusing to let me turn away as he wiped my tears. When I told him, three months later, that I couldn't return to the classroom, he encouraged me to pursue my dreams, whatever they might be. I wasn't Marcus's prototype; I was the person he'd chosen. For once, I was the one. But being Marcus's person was no longer enough to satiate me.

I had to shake up my life to survive. I knew that, but I couldn't tell Marcus. Through it all, I worried most about disappointing him—and that was the problem I needed to solve. I didn't only need to escape teaching and escape Denver. I needed to escape Marcus. I didn't know who I was without him anymore. I was so ensnared in our love bubble, shaping myself to be whomever Marcus wanted, that I'd isolated all the other planets in my social system of healthy relationships. I hadn't communicated with my friends in months. I'd stopped making new friends, so Marcus was the only person I could rely on. I didn't have anyone I could turn to who would understand. As my therapist slowly helped me awaken to all these new realities about how I'd shrunk myself and what it would take to reclaim my identity, a bud of hope emerged: a digital media job in New York at a women's lifestyle website. It was my chance to begin again, to uproot my small family and move to a new city with new opportunities. Marcus and I created a plan: He'd stay in our apartment for a month after I left for New York, and I'd use that time to get us set up. I'd get settled at work, lease an apartment big enough for us and our two sixty-pound dogs, and begin scouting potential jobs for Marcus. None of that came to be.

It would take six or so months before I realized Marcus wasn't coming and it was time to settle into this new life alone. Instead of moving to New York as we'd planned, he decided to move to Michigan to become his terminally ill father's caretaker. As the distance began crumbling the foundation of our relationship, I slowly reclaimed the bits of my life I'd sacrificed to be with Marcus and feel loved by him. Instead of concerning myself with what pleased him, what he enjoyed eating, what he expected from me, I now had to live only for me. I subleased the brick-accented Brooklyn studio I'd always wanted to move into and decorated it in the exact way I desired. I began staying out late to enjoy happy hour with my coworkers, and lost myself in the city's library shelves for hours. While Marcus and I still FaceTimed every single day, his absence allowed my other relationships to bloom. I reconnected with the friends I'd been neglecting, started making and keeping brunch and dinner plans, and traveled wherever my heart desired.

I was still devoted to Marcus; we were still in a committed

relationship, so I didn't entertain other men, but I entertained myself. I rediscovered what gave my body pleasure, and I masturbated until my touch felt as if it were enough again. I didn't need anybody else; I had myself. Thirteen months after I'd moved to New York, Marcus came to visit—and his mere presence felt invasive. I enjoyed showing him the city's historic landmarks and holding his hand on the subway, but welcoming him into the apartment I'd cultivated for myself just didn't give me the jolt of joy I'd thought it would. As soon as he came through the door, I was ready for him to go. My life was my own again. I no longer wanted to share it with Marcus or anyone else. Six months later, Marcus and I decided to end our relationship, and it felt like a boulder had been lifted and I could *finally* breathe again. Whereas breaking up would've felt devastating a year before, now it felt like the perfect next right step. I'd been the prototype and then I'd been chosen, but neither fulfilled me as much as being alone on my own terms. Being single, then, became comfortable and almost like a lucky charm. The world felt as if it expanded as soon as I decided to leave Marcus behind. My career blossomed. My rekindled friendships became life-sustaining. My emotional and mental health were no longer in flux because I was providing my own stability. For nearly three years, I settled into my existence as a single woman concerned only with my own happiness. It felt fulfilling. When people would ask me when I was going to put myself out there and begin dating again, I'd brush them off. I didn't need a partner. I had me.

Only one thing could have made me reconsider the role

of relationships in my life: Queen Latifah rom-coms. First, I watched 2006's *Last Holiday*, where the Queen stars as Georgia Byrd, an introverted saleswoman who refuses to take risks. While she cooks gourmet meals for one of her teenage neighbors, she eats Lean Cuisine for dinner. She has a crush on her coworker Sean (LL Cool J), but she keeps buying barbecue grills and other unnecessary appliances from his section instead of asking him out. But when she's mistakenly diagnosed with a terminal illness and given mere weeks to live, she cashes in her savings to embark on a once-in-a-lifetime European vacation.

By stepping out of her comfort zone, Georgia's able to reclaim the parts of her life that she never felt were open to her. When Sean learns about her diagnosis, he follows her to a luxury resort in the Czech Republic and expresses his love to her—just as she learns she's not dying after all. Georgia and Sean live happily ever after, but, more important, Georgia gets to pursue her dream of opening a restaurant with her husband's support. *Last Holiday* hit so me hard that I immediately watched 2010's *Just Wright*, in which Queen Latifah brings downtrodden physical therapist Leslie to life. After her godsister, Morgan (Paula Patton), begins dating professional basketball player Scott McKnight (Common) and he gets injured, Leslie becomes his live-in therapist. When Morgan dumps Scott because of his injury, he slowly realizes that Leslie is the only one for him. To put it lightly, the movie is corny as hell, but Leslie's career ambitions and unwillingness to play second fiddle to Morgan—even dumping Scott when he can't choose between them—made my chest fill with possibility.

Could I date again? And, if I did, could I do as Queen Latifah's characters did? Would it be possible for me to prioritize myself as I dated? Yes. I'd spent three years relearning myself. I was one year into my heart failure journey, and I'd slowed down my life to focus on my health. It was the perfect time to consider the realm of possibility of welcoming a new man into my life. This time, I wouldn't shrink myself and I would end relationships that didn't serve me. I wouldn't let another Marcus consume me. I'd stay focused on my health and my career. I'd prioritize my friends. I'd still be me, even within a partnership.

I thought I'd found the perfect partner in Dominic, but he was too emotionally unstable. When he kept putting an unloaded gun to his head during our FaceTime calls, I ended our relationship and encouraged him to seek therapy. I then met Shakeem, a recent divorcé whom I shared so many interests with. We spent hours talking about the books we loved and his career as an activist, but when I learned that he wanted to have a polyamorous relationship, I immediately stopped sleeping with him and offered friendship instead. As much as I enjoyed spending time with Shakeem, we didn't have the same intention for our relationship, so I gave him what I could offer—an ear and a shoulder when he needed it. And then there was Kingston, a single father with eyes so beautiful that it always felt as if he was glamouring me when we woke up together in the morning. I loved him for nine months. I loved him deep. Our bond was so uniquely strong that I thought we'd end up together. But when I learned that he had a dismissive, avoidant attachment style and he was incapable of effectively communicating during conflict, I ended that relationship, too.

I refuse to settle. It has taken me a long time to both utter that sentence out loud and use it as a guiding principle when pursuing potential relationships, most likely because I've feared that having such strict boundaries about finding an equal will leave me alone in the long run. It would be nice to find a partner, but I'd much rather remain single than settle for someone who's undeserving. Black women, especially fat Black women, however, are often given the message that it is better to be paired with someone, anyone, than to navigate life alone. Even if we have to perform the majority of the housework and child-rearing while also bringing in most of the income, those obstacles are better than going it alone. We're sold the twisted and unhealthy idea that it is better to be chosen, even when the choice isn't explicitly ours to make, than to be overlooked or considered unworthy of romantic companionship.

In the long-running CBS show *Criminal Minds*, Penelope Garcia, played by the inimitable Kirsten Vangsness, brings the party to the FBI's Behavior Analysis Unit (BAU). While her colleagues wear bland black outfits, Penelope dons weird, color-coded ensembles that always make her stand out. In the third season, Penelope meets James Colby Baylor (Bailey Chase), a charming and handsome Romeo who insists on taking her on a date. BAU profiler Derek Morgan (Shemar Moore), with whom Penelope spends season after season flirting, tries to warn her that there's something odd about James, and that she shouldn't go on a date with him. But Penelope mistakes Derek's warning for a jealous slight. Because of her size, she seemingly thinks that there's no way a man as successful and attractive as James—or Derek, for that matter—would be interested in her without wanting to take advantage of her.

Penelope decides to go on the date with James, and at the end of it, James shoots her in the chest and makes it look like a botched robbery attempt. We are to believe that Penelope, an outwardly flamboyant former hacker who wears the brightest, wackiest outfits and openly flirts with the sexiest member of the BAU, is gullible enough to fall for this serial killer's ruse. She can't use the profiling instincts she's cultivated for at least two seasons to detect that the man who's pretending to be interested in her only wants to harm her.

The episode is framed around Penelope's supposed naïveté. She doesn't realize that James's white four-door sedan is more than likely rented, or that he's not actually a lawyer, though he uses legal jargon. She tells Derek that she just thought he was nervous before their good-night kiss, instead of anxious about the crime he was about to commit. No matter how confident Penelope appears to be, the mere thought of gaining attention from a potential romantic partner makes her lose her sense of self and ignore that small voice that tells her something or someone isn't right for her. BAU profilers ultimately capture James, who was using an alias, but the message is still loud and clear.

I want myself too much to ever put myself through such torment again. You see, being the prototype devastated me over and over again, so when I was finally taken off the floor and taken home, I just wanted to hold on to my partner by any means necessary. But through this journey I realized that the only person I really need to hold on to is myself. Eight months into a new relationship with a completely new dynamic, I finally figured it out: no man could make me a prototype when I've already chosen myself.

CELEBRITIES, WEIGHT LOSS, AND US

I feel as if my body is withering away. I don't say this as a form of hyperbole or in a quest for sympathy from strangers. It's the rotten-bone truth: Since being diagnosed with heart failure in 2019, my body has considerably shrunk in size—to the point that I've had to discard my entire wardrobe and purchase a new one. That's not a compliment. This isn't earned or intentional weight loss. I wasn't aiming to look in the mirror and see a smaller face or slimmer waist staring back at me, which makes the fact that it has happened even more jarring. It often feels as if I'm looking at someone else, someone chiseling out because of illness, but for those surveying my body—strangers and loved ones alike—my newfound body is worthy of being praised, no matter how the weight loss was achieved.

"You look amazing" is a common refrain, as is "How much more weight do you plan to lose? You look great!" I've been told I'm glowing so often that it feels like my face must have been permanently smeared with bronzer, just enough to keep the compliments flowing. Though their intentions aren't malicious—effervescent praise rarely is—such praise reinforces the very system that I've been attempting to divorce myself from for nearly a decade now. Thanks to a culture that is equal parts obsessed with diet culture and consumed by the thin ideal, we're all supposed to be on a never-ending journey to conform to the standard, and when we're able to achieve that—even if it comes through progressive, chronic illnesses, as it has in my case—it is worthy of both reward and praise. From the perspective of those issuing these compliments, even if weight loss is the consequence of debilitating illnesses that could ultimately kill me, the sickness was worth it.

No one cares why you've lost weight. Whether it's accidental, intentional, or a by-product of sickness, we're taught that a slimmer body is to be marveled at—a sign of a newfound discipline and an adherence to the thin ideal after being disruptive to the social order for so long. Whenever I discuss what heart failure has done and continues to do my body—exhausting me without warning, stripping me of the Friday-night baths I used to look forward to, and causing extremely painful cramps to roil through my legs and feet—my feelings are cast aside as people gush about how good I look. "You're beautiful now" is a common refrain. "You're so small" is another. What I hear is: heart failure might have cost you, but sickness has also granted you something more important than your aches and pains.

To be clear, I'm not supremely thin. Nobody would mistake me for skinny. My body is still fat, but it's considerably smaller than it used to be. My face is smaller. My legs are smaller. My old clothes hang off my frame as if they once belonged to someone else. While this may seem like a point of pride for those who look adoringly at me through fresh eyes, impressed by what I've "achieved," it is a point of contention for me, a point of sadness, and, in some respects, a point of no return. When I look in the mirror, I don't see a body slimmed through strict dieting and weight loss. I see a body that's weary, that's battered, that has been through absolute hell. I see a body that's resilient and has gone through the wringer to keep me alive. None of what I've endured matters in this un-winnable scheme: I am thinner, and therefore everything I've experienced to get here is secondary. It's an experience that's common for fat women, especially those in the public eye who've lost weight, and it's reflective of the ways in which we understand (and don't) a fat person's relationship to thinness.

When Adele debuted a thinner body on Instagram in 2020, she was bombarded with compliments from normal people and celebrities alike. Though the caption of her photo acknowledged the sacrifices of essential workers during the pandemic, her body became the hyperfocus of the commentary. "YOU LOOK AMAZING," YouTuber James Charles commented, while Chrissy Teigen followed with "I mean are you kidding me." Lil Nas X, an Instagram account called Healthy Fitness Meals, Jenni Farley, a.k.a. JWOWW, and a mix of other ordinary people and celebrities descended on this picture to praise Adele. One comment, from Nagore Robles, immediately grabbed my attention because it summed up the issue far better than I ever could have: "You've worked so hard to look like this and I'm so glad you've got your purpose, but I want to tell you that for me you were always a spectacular, beautiful, and sexy woman."

Few people, if any, mentioned Adele's divorce, which she attempted to handle privately but was still forced to mediate publicly as tabloids speculated about everything from whether she and her ex-husband signed a prenuptial agreement to how much she'd have to forfeit in alimony to how they'd navigate custody of their small child. Divorce takes an enormous toll on the majority of people who experience it. For many, it can resemble a death—the forever separation of two lives that were once melded together—and that can result in both emotional and mental turmoil for both parties. Given what we know about what divorce does to regular folks, why would people praise Adele for being smaller in the aftermath of hers? "I had the most terrifying anxiety attacks after I left my marriage," Adele told Oprah Winfrey during an interview in 2021. "They paralyzed me completely and made me so confused because I wouldn't be able to have any control over my body." Working out, then, became a means of improving her mental state because she trusted her trainer and the exercise routine gave her life more structure. As a result of this life shift, she lost 100 pounds. To presume that she's smaller because she intended to be is to presume that all people who lose weight do so intentionally, only reinforcing the very surveillance that often precipitates fat people pining to lose weight to begin with.

When you're fat, every decision you make—or don't—becomes subject to the scrutiny of strangers. If you eat too much or too little, it becomes worthy of commentary. If you lose or gain weight, your body is under a microscope, one that's further magnified if you are a woman or a celebrity. After all, tabloids devoted entire issues to celebrities' "best and worst bodies" until the practice fell out of favor in recent years. But fat celebrities, in particular, are subjected to an even more gross invasion of their privacy. Whether it's Kirstie Alley (politics aside) and Adele or Mo'Nique and Oprah Winfrey, there's a cultural investment in how much their bodies lose and gain and what has prompted it.

In the case of Adele, this is largely pure speculation. Despite never discussing her body publicly, she has still become ground zero for people to plant their flags on, to project a meaning she never articulated onto her weight loss. What Adele chooses to do with her body is her business. How we perceive her body, however, reveals all our cards: Our culture treats weight loss as if it's about aesthetics or a vain quest to assimilate. But we rarely consider that it is sometimes a consequence of traumatic experiences.

When comedian and actor Mo'Nique first began her career in the late 1990s and early aughts, she prided herself on her provocativeness, her boldness both on and off the stage, and on being a fat stand-up comedian who was not ashamed of the size of her body. Her tagline was often "I'm not afraid of you skinny bitches!," and she'd loudly cackle about skinnier women needing to eat more and put on some weight in order to compete with her. Mo'Nique's confidence radiates whether she's performing in a sold-out stadium, as she did in *The Queens of Comedy*, or at a prison, as she did in a stand-up special. She has remained unfiltered about everything from her open relationship with her husband to the ways in which the television and film industries malign and attempt to low-ball Black women.

That honesty also translated into her brazen openness about the size of her body. After winning the Best Supporting Actress Oscar for her role in *Precious*, Mo'Nique began losing weight—and documenting the experience online. She used social media to chronicle her weight loss, which she achieved through a raw food diet as well as dance-oriented and cardio-focused exercise, and she was open about her reasons for losing weight. In an interview with late-night host Arsenio Hall, Mo'Nique said, "When my husband asked me my weight, I answered and he said, 'That's too much weight. I want you around for a lifetime and that's not healthy.' It was at that moment that I went through guilt, I went through shame, because of my size. Because I never felt love like that before." She had developed high blood pressure, and she realized that she wanted to be around to raise her two children, who were both under ten at the time. "I want to meet their children. I want to be able to play with their children. I don't want to be a burden on my family due to self-neglect. . . . I was fortunate to watch my grandmother play with my children. I want to be in the same position."

Mo'Nique chose to lose weight to "save her life" because she thought the size of her body was endangering her health, which is a stance complicated by the presence of fatphobia in medicine, but nonetheless one that spoke clearly to the personal, intentional nature of her decision. She never once mentioned that she'd lost weight to fit into a sample size for a red carpet appearance or because she'd felt as if she would be more beautiful if she were thinner. She lost weight for a reason that felt important to her, but somehow, all we could culturally latch on to was the fact that she lost weight at all. Headlines about Mo'Nique blared about the exact number she lost, her goal weight, and how she achieved her weight loss, but few explored the reason she was smaller—because it doesn't matter in the thin imagination. Her conformity is enough, regardless of her reasons.

It's the same experience I've endured as I've shrunk into a woman I can no longer recognize. I would be content with my fatter body if that meant I wasn't in heart failure, if pulmonary hypertension hadn't stripped me of my ability to experience pregnancy and to birth my own children. I would be content with my fatter body because it was mine, not forced into submission through a strict sodium and liquid regimen that often leaves me feeling supremely dehydrated. I would be content with my fatter body because I *was* content with my fatter body. It's beyond time to undo our cultural assumption that weight loss is a triumph. When someone's body has changed, we still can't seem to keep ourselves from gawking and speculating and discussing. We rarely stop to consider why other people's bodies should be so fascinating to us at all—and what might change if we simply minded our body's business and thought about other people's less.

OUR 600 LB. OBSESSION

*E*ach episode of TLC's long-running docuseries *My 600 lb. Life* begins in the same fashion: with a grunt and a sigh. *My 600 lb. Life* purports to be the standard reality television series, no different than *The Real World* or the *Love & Hip Hop* franchise, except the subjects of the camera's intense gaze are people who weigh more than 600 pounds and, in extreme cases, close to 1,000 pounds. Across both elements of the franchise—*My 600 lb. Life* and *My 600 lb. Life: Where Are They Now?*—we're introduced to people classified as "morbidly obese," meaning they're 100 pounds or more above their "ideal" body weight, without regard for how that designation is used to stigmatize people. For one year—two years if they're featured on *Where Are They Now?*—we follow each patient as they're guided through a weight-loss program created by Dr. Younan Nowzaradan, a surgeon who behaves as if he's the Gordon Ramsay of weight loss.

Nowzaradan is known for being blunt (and calling it

being honest), acting in ways that can seem incredibly cruel, and literally telling his patients they're one extra cheeseburger away from becoming immobile—if they're not already—and dropping dead. His high-protein, low-carb diet program, which encourages patients to consume 1,200 calories a day, is one of the few in the United States that focuses specifically on people for whom it's deemed too risky to have bariatric surgery because of their size. Dr. Now, as he's often called, requires his patients to lose a certain amount of weight before performing a gastric sleeve or full gastric bypass to ensure the surgery is as safe as possible and they have less chance of dying on the operating table.

Over the course of two hours of television designed to humiliate, we follow each patient's journey—beginning with said fat person, who is identified by their first name, age, location, and approximate weight, typically struggling to get out of bed, use the bathroom, and shower. Sometimes, the subject—not a person deserving of dignity, but a subject—can walk to the bathroom on their own. But often, a caretaker—their child, a neighbor, a romantic partner, a next-door neighbor—helps them sit up, helps them walk to a portable toilet, and then helps them into the shower. Sometimes, they have to literally shit in their beds because they've completely lost their ability to walk. Once they're in the shower, the only thing preventing their most intimate parts from being displayed is a small blur similar to the one TV shows use to cover a raised middle finger.

Subjects who are able to shower treat us to a close-up: soap pouring over their body as they attempt to clean between every fold on their flesh because, as many of them say, their bodies may develop an infection if they're not immaculately clean. Once that's done, it's on to the most important thing: eating. The camera zooms in as the subject eats whole slabs of bacon, boxes of doughnuts, multiple pizzas, fast food, fast food, and more fast food. From the show's perspective, we have to see these subjects eating because that's the only way the disgust will *really* sink in. *These are fat people!* the show tells us. *Look at them! Do you want to be like them? Stop eating!*

It's at this point that the show fully loses itself to depravity. The subject details their entire life story from childhood, mostly focusing on how and why they've gained weight. The subjects were once, as they often put it, "normal size"—until something traumatic caused them to begin gaining weight. The show reserves between three and five minutes for them to document their worst experiences—incest, rape, the untimely death of a parent, being rejected for being transgender—and then conveniently tucks those tears away, because that trauma is deemed a distraction from the overall point of the show. Yes, their pasts are sad and all, but we're here to poke fun at these fat people, and we shouldn't ever forget it.

By the time our subjects arrive in Houston to meet with Dr. Now, they've all declared that they're at their wit's end. They're going to die if they don't lose weight, and he's their last hope before they eat themselves into an early grave. Dr. Now is sure to reiterate that—over and over again—as the subjects break down in front of him, begging for his help and saying that they have no idea how they got to the point of being bedridden. He doesn't peel back this layer of the onion either—being fat is enough. We don't see him sending his patients to psychotherapy to address their issues or immediately referring them to a professional nutritionist who can help them learn how to nourish their bodies without "overeating." Instead, he gives them a weight-loss goal—lose 50 pounds, 60 pounds, or 70 pounds in the next two or three months—and a strict diet. Eat 1,200 calories, walk every day to strengthen your body, and you'll be good as new, mind and emotions be damned.

Following the success of *My 600 lb. Life*, TLC released *Too Large*, another reality TV show about fat people losing weight, in 2021. Though it has the same premise as *My 600 lb. Life*, it isn't as degrading, because Dr. Charles Procter, the bariatric surgeon who oversees the show's weight-loss program, helps the patients in every aspect of their lives instead of berating them for ratings. He recognizes that many of his patients—like Corey, who gained weight after his mother rejected him because he's gay—need more than a strict food regimen. They need a doctor who cares about their overall well-being. While Dr. Now typically doesn't send his patients to psychotherapy until they're in danger of regaining the weight they've lost, Dr. Procter immediately refers his patients to a counselor. While Dr. Now pokes fun at his patients who are struggling with their hygiene, Dr. Procter talks to his patients about the benefits of hiring an in-home health aide. While both shows treat fatness as a humiliating condition, *Too Large* has found a way to humanize the people it follows, allowing them to keep their dignity. It's an exception in a larger pool of shows that would rather embarrass fat people than help them in whatever way they need, whether it's losing weight or getting a housing voucher to prevent them from becoming homeless.

There comes a time within these shows when the subject is encouraged to "debut" their smaller body, showing off their new figure to family members and friends who've been waiting for a reveal that proves their loved one has been cured. These previously bedridden people are now able to fit into clothes that used to be inaccessible to them. They're off crutches, out of wheelchairs, and able to walk around their neighborhood. Their children respect them again because they no longer have to take care of them. They start dating and, sometimes, they fall in love. They become visibly confident, wearing flashier, splashier shirts and bolder wigs, putting on a piece of jewelry or two, adding an extra switch to their walk. These shows tell us that losing weight not only lengthens a person's life but also gives them the confidence to face life's obstacles, take a risk on love, and make others proud in the process. What they don't tell us, however, is the moral conundrum that accompanies shows airing some people's last moments for the sole purpose of poking fun at them. Multiple patients have died either while filming *My 600 lb. Life* or in the months and years afterward, and their deaths are nearly always attributed to "obesity." Many of them die from heart attacks, a fear that plagues me as someone with heart failure, so Dr. Now is often praised in Reddit forums for attempting to save their lives. But what kind of savior would flaunt a person's sickness on television for ratings? Apparently, one who understands that fat jokes are a specific kind of currency that can always be cashed in season after season.

MY MOTHER AND I OFTEN USED TO JOKE THAT WE SHOULD ATTEMPT TO BE cast on *The Biggest Loser*, especially when the show was in its heyday on NBC.

"We should throw our name in the ring," my mother would say to me half jokingly.

"Oh no, I don't think so! You won't embarrass me on television. Plus, we're not even fat enough. They'd kick us off in two weeks."

The Biggest Loser is a reality TV show that rewards people for losing weight. The person who sheds the most pounds, no matter how it's attained or if it's sustained after the season ends, is awarded a cash prize and the adoration of millions of television viewers who just can't believe how quickly they dropped *all* those pounds. For several years, *People* magazine would dedicate a cover to debuting the svelte new body of each season's winner. During the show, each contestant would discuss how their weight was holding them back in various elements of their life—from their ability to find and retain gainful employment to their mental health. Once the number on that scale is lower, the world is suddenly more open than it has ever been before.

Their weight—their "super obesity," if that phrase floats your boat—is preventing them from living even a marginal portion of the lives that should be open and available to them.

But in order to get access to that brand-new life, there's a grueling gauntlet fat people must endure. We must push our bodies to their limits on national television to show the normal, thinner world how miserable we are in our current states. We must prostrate ourselves in front of those who can help us see the error in our ways—in this case, the bariatric surgeons and fitness coaches who will scream at us until we see the light. And then, we must restrict how much food we consume and exercise excessively until we lose enough weight to qualify for the cover of *People* or for bariatric surgery.

For many years, I avoided shows like *My 600 lb. Life*. Though my mother, who has experimented with nearly every diet on the market, was a devoted viewer of what I've referred to as the fat reality TV show vortex, I'd leave the room when *The Biggest Loser* or *True Life: I'm Fat as Hell* graced our screen. I did this for myriad reasons, but let's start with the obvious. These shows are ridiculously exploitative. People lose and gain weight all the time, but shows that use this premise to drive ratings—whether it's WE tv's *Mama June: From Not to Hot* or TLC's Wednesday slate of fat-focused television—reinforce the very idea that fat bodies are a problem to be solved, a conundrum in dire need of being fixed. Even the milder fare, like *1000-lb Sisters* and *One Ton Family*, which are exactly what their titles suggest, feed into this idea. Eventually, though—as is often the case when you're living in the same home with someone—my mother's consumption interests bled into my own. *My 600 lb. Life* became one of the shows that we shared, that we are within the universe of, that we have a secret language about. "Have you watched our show?" became second nature for us on Wednesday nights after we'd both finished working.

My obsession was sparked by a single episode: A forty-two-year-old Black woman named Octavia lives in Kansas City, Missouri, with a roommate. She weighs more than 650 pounds and hasn't left her bed in nearly three years, so she uses a pail at the end of her bed as a toilet. Her roommate, who moved in with her when she became bedridden, brings her a bucket with soapy warm water so that she can wash as much of her body as possible. The best moment in Octavia's entire day is when her sister brings her niece and nephew to visit her. Their visits increase her social interaction and allow her to pressure her sister to purchase her lunch and dinner. Candy is the language she shares with her niece and nephew, whom she allows to reach into a huge plastic bag and remove a piece or two to enjoy. Though Dr. Now tells Octavia that it's dangerous for her to move to Houston to be near his clinic, she disobeys him, packs up her apartment—her bedroom had been in the living room to begin with—and rides in the back of a van, a trip so painful that she complains about her aches and pains the entire hours-long ride. When she arrives in Houston, she's immediately chided by Dr. Now for making the trip, and then shot down. Because Octavia hasn't lost any weight since their initial phone call, she'll need to lose more than 50 pounds in two months in order to qualify for bariatric surgery.

Dr. Now is worried about Octavia's overall health because she's unable to stand or walk, so he hospitalizes her to speed up her weight loss and help her begin walking. Though not every patient is hospitalized, many of them are, depending on the overall state of their health. Octavia's put on an intensely restrictive diet, and nursed—tough love, be damned—back to some semblance of health. By the time she leaves the hospital, she's lost more than 60 pounds and she's able to stand for a short period of time and take a couple of steps. It isn't enough for bariatric surgery, but it's enough to be released from the hospital and prove to Dr. Now that she can lose weight on her own. But when she gets home, she gains weight instead. When she comes to the clinic for her next weigh-in appointment, Dr. Now tells her she's always blaming someone or something for her weight gain instead of taking control of her health. In his eyes, Octavia's lack of discipline is the problem and it's her lack of discipline that will eventually kill her.

As I watched the episode, I felt frustrated by Octavia. Why couldn't she just say no to the "unhealthy" foods her roommate is bringing into their apartment? Why can't she just force herself to walk, even if it's just from her bed to her front door and back? Though I know weight loss is never as easy as doctors purport it to be, I still bought into Dr. Now's manipulative rhetoric. After all, *lose weight, save your life* doesn't seem like that difficult of a mandate. I can remember turning to my mom on the couch and asking, "Does she want to live or not?" She just shook her head.

By the end of that episode, I was hooked. I went back and watched every season of the show, and I kept cycling through those same emotions: jubilation for those who follow Dr. Now's program and lose weight and frustration with those who just can't seem to get themselves together enough to follow his diet. For those seeking Dr. Now's care, eating "healthy" and exercising isn't enough. I know that. And yet, watching fat people try and fail to lose weight still ensnares me so much that *My 600 lb. Life* has become appointment television for me.

I've been studying fatphobia for so long that I can pinpoint obscure fatphobia in the most "body-positive" pop culture. I know *My 600 lb. Life* shouldn't be airing on television. I know TLC has built its wealth on the literal backs of people from marginalized communities, including trans people (*My Pregnant Husband*; *I Am Jazz*), little people (*Little People, Big World*; *The Little Couple*; and *7 Little Johnstons*), undocumented immigrants (*90 Day Fiancé*, anybody?!), and fat people, but somehow, a network that puts a bull's-eye on the back of the very community I write and advocate for draws me in week after week.

As ashamed as I am to admit this, there's a particular two-part episode of *My 600 lb. Life* that I've watched, repeatedly, both on TLC and in perpetual reruns on Hulu. It features Janine, a woman in her midfifties who is only able to transport herself by motorized scooter from her bed to her bathroom and to a local McDonald's and back. There's no doubt Janine is lonely, and she repeatedly says that she's miserable in her body. She lives alone in Seattle, hundreds of miles from her siblings, who reside in Colorado. Both of her parents, who adopted her when she was a few months old, are deceased, and she ponders if they ever really loved her to begin with. She believes they intentionally treated her differently than her siblings. Knowing that Janine's biological mother was fat, her adoptive mother put her on a diet at the age of four, and she's been on the wretched dieting path ever since. Janine is so lonely, in fact, that her cat is her only company, and she carries on full conversations with him as she eats waffles covered in strawberries and whipped cream.

If *My 600 lb. Life* were actually invested in the health of its subjects, Janine's tormented past—being abandoned, being adopted into a family that cared far more about the physical size of her body than her mental health, and being so isolated that she has to hire a stranger from the internet to travel with her to Houston—would've been Dr. Now's primary focus. Instead, Dr. Now repeatedly shames Janine for not seeking care sooner. It begins when she attempts to travel to Houston by plane and suffers a debilitating panic attack that forces her to be carried off the plane on a stretcher and hospitalized. Next, Dr. Now yells at her for not abandoning her motorized scooter, though it's clear that she's unable to walk without it. When she begins telling Dr. Now's nursing staff that she can't walk without assistance, he calls her "delusional." We see, repeatedly, that Janine is calling out for help, and no one is willing to step in and intervene in a way that preserves her dignity. It takes two hour-long episodes, multiple breakdowns, and an invasive procedure that literally causes her hair to fall out for Janine to be taken seriously enough to meet with a psychotherapist, who simply tells her to forgive herself for everything she's done to "punish" her body.

I have watched this episode of *My 600 lb. Life* so often that I can recite it nearly word for word, starting with the moment Janine uses a metal contraption of sorts to wipe her butt and says, with jest and in full view of the camera, "Want to know how a fat woman wipes her ass? This is how." There's no logical reason for me to be so obsessed with this show—even browsing Reddit threads during and after episodes to gauge what other people are saying, and crying when episodes end with the subjects dying. But as I've watched and laughed and cried, it's become clear to me: feelings of superiority can breed the most monstrous behavior. Fatness is intertwined in my very being; I don't know who I am if I'm not fat. But I've also prided myself on not being *that* fat.

Many fat people have that moment when we walk into a room—sometimes with strangers, sometimes with friends—and we look around, sizing up everyone around us and attempting to figure out if we're the largest person there. For me, that moment dictates a lot about what happens next. If I'm the largest person in the room, I feel an innate urge to shrink, to make myself as unobtrusive as possible. I position myself in a corner where nobody can see me, lean against the wall, and pretend as if this is natural posturing. But on the rare occasion when I'm not the largest person in the room, when I size everyone up and gleefully notice someone bigger than me, it imbues me with a wicked, false confidence: "Now there's someone else to focus their attention on." I no longer have to be a wallflower because someone else can fill that role now. In these moments, I siphon that other fat person's energy. Suddenly, I have an effervescent buoyancy, and I feel confident enough to socialize, to speak to strangers, to put my shoulders back, hold my head high, and own a room.

That same dynamic functions in my obsessive, repetitive watching of *My 600 lb. Life*. As I absorb the subjects' stories, no matter how tragic they are, I get to reinforce to myself that I'm superior because my weight is "under control." I'm still mobile. I can still walk into a plus-size store and purchase clothes. Doctors discriminate against me, but I'm not at the mercy of any particular provider to receive the care I need. In other words, I'm able to create distance between the fat body I inhabit and their fat bodies. I know I gain nothing from investing in fatphobia and perpetuating it toward those larger than me, and yet, it's much easier to laugh at fat people on television than to think about those laughing at me.

Dr. Now routinely tells his show's subjects that they smell, that their bodies have an odor, that they could be living a better life if they just slimmed down. He often couches these insults in humor and feigned concern, pretending as if his cruelty is *really* about preserving his patient's life. The reality, however, is that he's able to play up their shortcomings for the camera—bringing in equal amounts of ratings and disgust from viewers who have bought into the idea that it's only their eating habits and never their genes or their trauma that have brought them to the point of immobility.

It's painful to admit that I feel morally superior to fellow fat people on reality television. I am sure—nearly positive—that this will disqualify me as a fat-positive activist. People might screenshot parts of this essay and spread it on social media to prove that I'm a fraud or that I purport to care about fat people in public, but in private, I am just as complicit in fatphobia as the very people and institutions I criticize. Unlearning is a difficult process. It first requires you to look in the mirror, admit that you've benefited from a system at the expense of other, more marginalized groups, and then actively

work to create new commitments and behaviors that dismantle that system. But when you've been indoctrinated into a
fatphobic theology where thinness is the god to be idolized,
and every element of your life underscores this worldview, it
becomes easy to pick apart people whose bodies are more unacceptable than yours, even if only slightly.

Peering out the window and asking, "Do I walk like her?
I hope not. She's waddling," covers my own fear about being
the person being judged in the way I'm judging. I deploy these
skills, forged in fire and struck against iron, while watching
My 600 lb. Life. I'm especially guilty of doing this when said
person refuses to follow Dr. Now's program and continues to
either gain weight or lose it more slowly than Dr. Now would
prefer. "That's an absolute shame" is one of my favorite retorts, followed by "Well, that episode was a waste of time.
They didn't lose any weight!" I couch these barbed comments
in the additional skin I've formed over time. I've learned that
it's better to strike first, to preemptively project before the
echo can return, even if it comes at the expense of your self-
image. You see, if I am obsessed with this show, then no one
can be obsessed with me.

YOUR LIFE IS DISPOSABLE

The week before the pandemic arrived on America's shores and permanently disrupted our lives, I was visiting Los Angeles for a weeklong leadership workshop for women in media. I should've appreciated that trip more. I didn't know—none of us could've known—the sheer trauma COVID-19 would begin inflicting on our lives in March 2020. If I'd known it would be the last time I'd stay in a hotel for thirteen months, I would've ordered more room service and stayed in those crisp white sheets just a little longer. If I'd known I wouldn't see my grandmother for fifteen months, I would've flown to North Carolina and hugged her until she shooed me away. If I'd known then what I know now, I would've tilted my head up at the sky and basked in the sun, let it shine brightly on my maskless face. I would've cherished the anxiety and fear I didn't feel, the absence of existential angst that then accompanied me to bed every night.

I'm not alone in these regrets, as millions of people have

died around the world, millions of others are grieving the unnecessary deaths of their loved ones, and millions of others are facing the possibility of never fully recovering from long COVID. We all wish we could've seen the pandemic bearing down on us and prepared ourselves accordingly, but hindsight can't save us now. By the time I flew home in mid-March, half of my fellow plane passengers were wearing N95 face masks, and, within a week, my entire state was shut down. Businesses were closed and schools were, too. Our local grocery stores were closing at 9 p.m., and everyone I know was transitioning to remote work. There were more FaceTime and Zoom calls and more overall scrolling, so much so that my phone's weekly usage report declared that I'd used it 63 percent more in the week I returned from Los Angeles. All these rapid changes to daily life felt disorienting, mainly because we were all learning in real time. There was a lot we didn't yet know about COVID-19 or the havoc the virus could wreak on our bodies. All we knew, all I knew, was that I wanted to avoid the virus at all costs. My cardiologist delivered that message loud and clear during an impromptu video appointment.

When he began the conversation with "Have you been outside? If so, don't go outside again," I knew there was cause for deep concern. I could feel panic burrowing itself under my skin and into my bone marrow. He offered up a simple reason for his directive: "COVID-19 appears to be impacting the lungs. It would make you extremely ill, and it may kill you. If you can stay home, please do. Don't fly. Don't go into grocery stores. Cancel all of your appointments." I nodded but remained silent, afraid that responding would make what he was saying reality. I didn't want to believe that my pulmonary hypertension put me so much at risk. In the weeks that followed, I completely adhered to his order. I didn't leave my home for a month. During that time, I felt like a fragile family heirloom that everyone was worried about dropping or tarnishing in some way. At first, the rules of quarantine were unclear, so close relatives were still coming to visit. Before people came into our house, they'd spray themselves with Lysol at the front door and immediately wash their hands in the hallway bathroom before speaking to me or hugging me.

When the Centers for Disease Control and Prevention assured us that COVID-19 isn't passed through surfaces in mid-April, I was finally able to leave the house for a walk. I didn't realize how much I'd missed fresh air until I felt the first cool breeze flow across my face. I don't have claustrophobia, but as a forever recovering agoraphobic, I worried deeply that staying in the house would retrigger that condition's appearance and alter the way I lived after the pandemic ended. With that fear in the back of my mind, I began driving my parents to the grocery store, though I couldn't go inside. I began walking every night, just to make sure I left the house. I thought COVID-19 was a temporary reality that we'd eventually be able to move to the other side of. I never could've imagined the pain, grief, and devastation the pandemic would bring.

As COVID-19 began spreading rapidly and claiming lives, it became very clear very quickly that the most marginalized among us—impoverished people, Black, Latinx, and Indigenous people, people with disabilities, and those with a combination of all these unique conditions that make them especially vulnerable—were most at risk of both contracting COVID and dying from it. Amid that chaos and fear, I turned to daily medical briefs for the most up-to-date information about the pandemic and whom it was striking.

Anthony Fauci, M.D., one of the country's top epidemiologists and infectious disease experts, who unexpectedly became a celebrity during the pandemic, would repeat, nearly every day, the demographic information of who was vulnerable to contracting a more severe form of the virus: people with compromised immune systems; elderly people, especially those in nursing homes; people with heart diseases, lung diseases, and diabetes; and, to throw a curveball, people with "severe obesity." Each time he included "obesity" amid these other chronic conditions, it felt like a punch to the gut. Several countries, including Great Britain and Mexico, began "cracking down" on obesity by banning junk food advertising before 9 p.m., requiring grocery store checkout lines to stock "healthier" foods, and banning stores from selling "junk food" to children.

"It's so clear that the overwhelming weight of serious disease and mortality is on those who are elderly and those with a serious comorbidity: heart disease, chronic lung disease, diabetes, obesity, respiratory difficulties," Fauci told *Fierce Healthcare* in March 2020. By including obesity alongside illnesses like heart disease and lung disease, Fauci subtly endorsed a controversial idea: that obesity is in fact a disease itself.

In June 2013, the American Medical Association (AMA) began recognizing obesity as a disease with "multiple pathophysiological aspects requiring a range of interventions to advance obesity treatment and prevention." The AMA wasn't the first organization to issue this guidance: the National Institutes of Health gave obesity the same designation in 1998, and the American Obesity Society followed suit in 2008. The medical community adopted people-first language—"person with obesity" rather than "obese person"—that separated a patient from their perceived illness. Doctors use an array of tools to diagnose obesity, from the misguided body mass index (BMI) to measuring "adiposity," or the amount of fat tissue within the body, to gauging a person's waist circumference. In accordance with other diseases, obesity was also broken down into classes of mild, moderate, and severe, depending on BMI.

This classification system opened medical avenues for "people with obesity" that weren't available before. Now, with obesity considered a serious disease, more health insurance companies have to either fully or partially cover the costs of weight-loss surgery and other medical treatments. Then the Internal Revenue Service classified treatment for the disease as a deductible medical expense, while the Social Security Administration began recognizing obesity as a valid source of impairment for disability claims. Beyond removing barriers to treating the condition, the AMA believed that classifying obesity as a disease would improve the public's overall understanding of it and remove the stigma that is often attached to people with obesity. Instead of obesity being viewed as a personal problem caused by bad choices, the AMA hoped to challenge weight bias in all its forms.

Rather than waiting for a person with obesity to develop

an additional illness, the medical community adopted a preventative approach to treat it as a metabolic disease that can be cured through diet, exercise, and if need be, metabolic weight-loss surgery. But this also opens risks for fat people to be treated *only* for obesity at the expense of their other ailments. When I first began seeing a new gynecologist in 2018 to treat a recurrence of fibroids, I expected her to focus on stopping the anemia-inducing bleeding and removing the benign tumors. As I sat on a tiny, mint-green exam table in a light-blue patient gown after my Pap smear and cervical biopsy, she instead told me she wanted to refer me to the clinic's metabolic weight-loss center. Losing between 50 and 100 pounds, she said, would help curb the effects of the fibroid and lessen the pressure in my uterus. I was crestfallen. I'd come for a solution, but I was instead diagnosed with obesity and directed to treat that nebulous ailment instead of the one acutely affecting my health. I just took the pamphlet she offered me and asked when I'd receive my biopsy results.

Around this time, the same doctor misdiagnosed me with polycystic ovary syndrome, a.k.a. PCOS. I didn't have any common PCOS symptoms, including follicles on the ovaries, excessive hair growth, or irregular periods, and blood work and an ultrasound didn't support a PCOS diagnosis. Still, she prescribed me Metformin, a medication commonly used to treat both PCOS and diabetes.

It would take more than two years for a reproductive endocrinologist, during the process of freezing my eggs, to definitively determine that I didn't have the condition. Deep down, I wonder if my gynecologist prescribed me Metformin not because she really believed I had PCOS, but to preventatively treat the diabetes she assumed I'd develop because of my weight. While the push to treat obesity as a disease does attempt to move the stigma from the people who have the illness onto the illness itself, it does little to address the reality that some people will always have larger bodies due to factors outside of their control—from genetics to lack of access to healthy food and lack of access to outdoor space to exercise—and that those people will need to be listened to and treated for other medical issues beyond their weight.

Designating obesity as a disease was supposed to address what has been called "the single greatest threat to public health for this century," by removing the idea that fatness is about lifestyle choices and treating it as a complex chronic illness. But it also gave doctors carte blanche to aggressively pursue obesity as the cause for all other illnesses, the single greatest disease in need of treatment, regardless of what other symptoms are impacting the patient and what they are seeking treatment for.

In March 2020, a neurologist tweeted (and later deleted) the idea that people with a BMI over 25 could be denied a ventilator because that's considered a "poor prognostic sign." In other words, a fat person's life wouldn't be worth attempting to save because they might die during intubation anyway. When I read the tweet, it filled me with a familiar sadness. Fat people were being deprioritized, as we always are. If we can't get our weight under control, we needed to be punished by doctors choosing whom to save based on their perceived worth. Soon, the hashtags #NoBodyIsDisposable and #NoICUgenics began trending on Twitter. While Fauci's goal in his constant mentions of obesity was to raise awareness about the risk factors for severe disease, instead, he contributed to fat people's panic about how they'd be treated if they contracted COVID-19. We had reason to fear that doctors would discriminate against us—and that our weight would be used to dictate the course of treatment.

For months after #NoBodyIsDisposable and #NoICUgenics trended, I still panicked every time I missed a step in my preparation to go outside. If I didn't sanitize my hands after coming in from my walk, I'd begin crying. I just knew I would contract COVID-19, need to be intubated, and be passed over for a thinner person on the mere basis of the appearance of our bodies. Though I'm prone to panic attacks, I had more of them from March 2020 to March 2021 than I'd ever had before.

When COVID-19 vaccines began rolling out in Colorado on a tiered system based on prioritized need, I was faced with an unfamiliar choice. I was thirty-two, a journalist—a group treated as essential workers within the state—with two chronic conditions, heart failure and pulmonary hypertension. But when it came time to register for the first vaccine, I was asked whether I had obesity before being asked if I specifically had chronic illnesses. I selected obesity as my preexisting condition. My cardiologist wrote a letter approving my vaccine because I'm immunocompromised; though the letter mentioned my chronic conditions, obesity was listed as the primary reason I needed to be prioritized for vaccination. My mom, who doesn't have additional illnesses, was prioritized for the same reason. My fat body was creating an inescapable conundrum: Do I adopt the "obese" label to skip the line and receive the vaccine? Or do I continue challenging the labeling of fat bodies as "obese" to begin with? I chose the former. I couldn't believe that being fat was benefiting me for the first time. It felt too good to be true, almost as if it was a loophole that shouldn't have been afforded to me. I carried a lot of guilt about getting vaccinated in the same class as my seventy-seven-year-old grandmother. I kept waiting for a nurse to side-eye me for jumping the queue.

Up until this point, being diagnosed with obesity had never been used in my favor. I've been punished for being fat. I've been punished by doctors. I've been punished by the school system. I've been punished by the dating-industrial complex. I've been punished and punished and punished, but now, amid a world-changing pandemic, being fat was suddenly an advantage. The guilt of that was suffocating, even if irrational. How could I be vaccinated before my parents? Why was the size of my body a factor in that? It's a conflict I'm all too familiar with: after being diagnosed with heart failure, I braced myself, time and time again, for a member of my care team to hint that I'd developed these conditions because I'm fat. But to my surprise, that has never happened. They said the exact opposite: I'd gained a significant amount of weight as a result of fluid building up all over my body, particularly pooling in my lungs and my ankles. Each doctor cautioned me about the dangers of gaining weight—because, for the first time, my weight gain implied that I had a worsening medical condition that required immediate intervention—and encouraged me to lose weight, as so many other doctors had done before.

The vaccine rollout was disorganized and uncoordinated, to say the least, with people navigating multiple vaccine waiting lists; calls to clinics, hospitals, and retail stores to inquire about vaccine availability; and the expectation to drop everything to head to a vaccination site at a moment's notice. But this might have been the first time that fat people were being treated with the care we've long deserved on a national scale. The term "obesity" has often been used to signal to fat people that we're incapable of taking care of ourselves. We have a mild, moderate, or severe condition, a ringing alarm that's supposed to signify to us that it's time to lose weight. Since we haven't heeded that call, we therefore can't be trusted to gauge what's happening with our bodies or make decisions about how our ailments should be treated. If you can't stop eating, if you can't exercise enough to burn enough calories to prevent you from gaining weight, if you can't learn how to control the intake of your food, then how can you determine if and when something is actually wrong with your body? Through this lens, only thin people have the authority to determine if and when their bodies are ill.

If fat people were treated as if they were a medical priority beyond just the COVID vaccine rollout, it could and would make all of the difference in the care we receive. When people are diagnosed with diseases like ovarian cancer or diabetes, there's a treatment plan created with the patient in mind and then tweaked as their illness either improves or progresses. When people are diagnosed with obesity, there is often no clear course of action. Instead, we're simply told to lose weight. We should restrict our calories, exercise more, and if all else fails, have some form of weight-loss surgery. There's no consideration about why we're fat, if our fatness is contributing to any health ailments we're facing or vice versa, or if a calorie- or carb-restricted diet and more exercise will actually result in weight loss. There is no mention of the fact that even if we do lose weight, most of us will regain it over time.

THERE'S A RUNNING JOKE AMONG FAT PEOPLE THAT IF YOU GO TO THE doctor for a sore throat, they're going to ask you to take a blood sugar test to make sure you're not diabetic. For many fat people, it is a harsh reality that any ailment whatsoever will be blamed on our fatness. Take, for instance, Eric Garner, the forty-three-year-old New York man who died after a Staten Island police officer, Daniel Pantaleo, placed him in an illegal choke hold. After Garner's murder, Pantaleo wasn't charged with a crime; the U.S. Department of Justice slow-walked an investigation into whether Pantaleo violated Garner's civil rights; and the New York Police Department allowed Pantaleo to remain on the force for five years, terminating his employment only after an internal administrative process ended in a suggestion that he lose his job.

During that hearing, Stuart London, a longtime police union attorney who defended Pantaleo, implied—quite explicitly—that Garner was responsible for his own death by virtue of being fat. "He died from being morbidly obese," London said. "He was a ticking time bomb that resisted arrest. If he was put in a bear hug, it would have been the same outcome." We're accustomed to victims of police violence being blamed for their own deaths—Michael Brown Jr. allegedly stole a cigarillo, while Philando Castile startled a police officer, and Breonna Taylor happened to be sleeping in her own bed during a no-knock warrant—and yet, it's stunning to see fatphobia, racism, and police violence converge to reveal the ways in which fat Black people are treated as if they're disposable. After Garner's death, Peter King, a House Republican who represented Long Island, said Garner died only because of his size. "The police had no reason to know he was in serious condition," King said. "You had a 350-pound person who was resisting arrest. The police were trying to bring him down as quickly as possible. If he had not had asthma and a heart condition and was so obese, almost definitely he would not have died." If Pantaleo hadn't choked Garner for selling loosies outside a store, he would still be alive. He wasn't a ticking time bomb who could drop dead at any moment; he was a man with a family who deserved to live. Fat people have long been treated as if we're disposable. That doesn't magically change because we were prioritized for vaccines. In fact, in many respects, it forced us to disassociate from our concerns about the faultiness of the BMI chart in order to preserve our lives and ensure our survival.

After waiting for two weeks, I received my vaccine. One month later, I received my second vaccine. Six months later, I got my booster, and I'll continue to get as many shots as I need to remain safe. I got to leave the house again. I'm traveling again. I got to hug my grandma and feel the sun on my face. The vaccine has been a relief, the one way to curb the panic attacks and give me permission to begin living again. If this is the one time fatness benefits us in a medical context, then we must take full advantage, but this shouldn't be the only time. We're all we got, and we have to continue to fight to be treated as fully human—not just a number on the scale.

BACK TO THE FAT FUTURE

For a long time, flying, or rather turbulence, created an unparalleled level of anxiety in me. It wasn't always this way. I started flying when I was five or six, my brother and I getting supreme "unaccompanied minors" treatment as we traveled from Denver to New York and back again. I was so comfortable flying, in fact, that I would fall asleep as soon as I plopped down in my assigned seat and strapped the seat belt across my waist. That all changed in 2014, when I moved to a rural part of Illinois for grad school. In order to get from that small town to the international airport in St. Louis, I'd board an eight-seater plane without window shades, a flight attendant, or bar cart service for a quick forty-five-minute sky trip through the Midwest.

When the plane hit turbulence during one flight from St. Louis to Marion, it felt as if we were going to drop out of the sky. As we hit bump after bump of rough air, my stomach felt as if it'd plopped into my lap, and I began hyperventilating.

The kind stranger sitting beside me gripped my hand and helped me breathe through the panic, but by the time we landed, I couldn't wait to spring from that aircraft and never board another one. I avoided flying for two years, until I wanted to travel with my friends and I remembered that you can't drive across oceans. When I decided to begin traveling by plane again, flying was all I could think about in the hours, days, and weeks before my trip, so to lessen the heart palpitations, sweaty palms, and overwhelming sense of impending doom, I created a routine that helped me manage the anxiety. I start by choosing flights equipped with wifi, so I can iMessage friends and family when anxiety strikes, picking a seat that's as close to the airplane's wings as possible, since that's allegedly the most stable area of the aircraft, downloading a movie or a few episodes of a TV show to watch, and selecting a compelling book that I can immerse myself in.

Though a fear of turbulence has given me a panic attack or two, this intricate routine has, over time, curbed nearly all my flying anxiety. But I've discovered there's no way to breathe through the judgmental stares of my fellow passengers whenever I ask for a seat belt extender, or to ignore the sighs, the eye rolls, and the obvious discomfort when I let someone know that I'm sitting beside them. It's almost impossible to feel comfortable in the tight space of a plane, especially when the seventeen-inch seats aren't built to accommodate a person with large hips or a wide waist. But the humiliation that comes with squeezing into an uncomfortable seat beside someone who'd rather sit beside a raccoon than share space with you only worsens the experience. The only time that I wish I were thinner is when I have to board a plane.

As writer Sesali Bowen put it in an article for *Refinery29*, "If I had a list of reasons to lose weight, 'travel comfortably via plane' would be at the top of it." That's how embarrassing and difficult flying is for plus-size people. Fat people are accustomed to being treated as burdens in tight spaces, such as airplanes, amusement-park rides, and even subway cars and taxis. Plus-size model and social-media influencer Natalie Hage learned this traumatic lesson when she flew from Dallas to Los Angeles in June 2017 for a modeling shoot. Once she got settled in her middle seat, Hage noticed that one of the people she was sitting next to was sending cruel text messages about her to one of his friends. "Hopefully she didn't have any Mexican food," the man's friend texted. "I think she ate a Mexican. I'm leaving a neck print on the window. If the news reports a DFW Airbus A321 leaving a runway without rotating, that would be my flight." Hage, who has more than 150,000 followers on Instagram and works with model Tess Holliday to oversee the #effyourbodystandards movement, decided to post about the experience on Facebook not only to blast this insensitive passenger, who she later learned was named Eric, but also to raise awareness about how fat people are treated on planes.

"I'm shaking right now," she wrote in her caption. "The texts were about me and I'm almost positive he took photos of me. Not only were the texts about me, but they were really mean and ugly, with even the recipient named 'linda' chiming back with shaming retorts . . . someone who can't even see the situation. I didn't do anything to him. . . . i am crumpled into a ball trying to not bother. I'm just so upset." Instead of swallowing that upset, Hage decided to confront Eric and tape their conversation as the flight landed.

"I couldn't help but notice that before we took off you were sending really horrible text messages about me to somebody," Hage said to Eric.

"No, I wasn't," he retorted.

"I have photos. Yes, you were," she continued, before reading some of the text messages she saw back to him. Eric apologized and blamed his behavior on being too inebriated. After Hage told Eric that his text messages made her uncomfortable, he went in for the kill: "In fairness, you probably shouldn't be sitting in an exit row seat. When they ask you, 'Are you willing and capable to assist people in getting out of the airplane in an emergency,' do you honestly think that you are?" Their tense exchange captures so much of what fat people endure when we dare to exist in a public space. "My body is none of your business," Hage rightfully responded. "Don't ever treat anyone like that again."

She's not the first fat person and won't be the last to encounter such mistreatment. I often pay additional money to preboard flights so I can get comfortable before the plane overflows. But that rarely keeps me from noticing the expressions of panic on other passengers' faces as they approach my seat, followed by noticeable relief when they realize they don't have to sit next to me—or resigned expressions when they realize they do. On one flight, the woman assigned to sit next to me sighed before she sat down, forced the armrest down, although it dug into my side, and smirked noticeably when I attempted to lower my tray table. Her unkindness made me uncomfortable, and so I, in the words of my grandmother, told her about herself. She apologized, of course, and said she was just extremely tired, but I knew the truth: she wanted to punish me for being fat, and she didn't expect me to rebuke her behavior and call her out. But these individual corrections are not, and will never be, enough to change fatphobia.

Discriminating against fat people isn't isolated to individuals, as Hage explained in her Instagram post: "This is a fat person's daily reality and not just on a plane. This is on a bus, standing in line at the grocery store, at a concert, on the internet. You can be completely in your own space, not bothering anyone, and people will still fuck with you and try to hurt you [and] all you can do is know you haven't done anything wrong just by existing and to move on." And when that's our reality, it's time to force a systemic shift.

Fat activist Stacy Bias captured the difficulty of flying as a plus-size person in her animated film *Flying While Fat*, which explores how the shrinking of airline seats to accommodate more passengers has affected plus-size customers. According to *USA Today*, legroom in coach has shrunk from thirty-five inches to thirty-one inches, and formerly eighteen-inch-wide seats have shrunk by an inch. Meanwhile, Americans are getting larger, and the Federal Aviation Administration is refusing to intervene by implementing rules that would create a minimum seat width and length. Instead, plus-size passengers are being subjected to different policies that cost us more money in exchange for baseline comfort.

For instance, Alaska Airlines, Allegiant, American Airlines, Frontier, Spirit Airlines, and United Airlines require fat passengers who can't put down both armrests and/or need a seat belt extender to purchase an extra seat. Rosie Mercado, a plus-size model, experienced this when she was traveling from Las Vegas to New York in June 2011. A flight attendant told Mercado she needed to purchase two fares before she could board the flight, because she was unable to fit in a single seat. Mercado told the *New York Post* in 2015 that the negative interaction was a major factor in her weight loss of more than 200 pounds. While it's understandable that airlines want to guarantee that all passengers are comfortable, why does that fall at the feet of fat customers, further increasing embarrassment and making travel even more unaffordable?

In Tyler Perry's 2007 movie *Why Did I Get Married?*, married couple Sheila (Jill Scott) and Mike (Richard T. Jones) are flying to Colorado to participate in their friends' annual couples trip. It's supposed to help the couples remember the reason they got married and reignite the romance in their relationships, but the other couples don't know Mike wants to use this trip to embarrass Sheila into asking him for a divorce. When Sheila can't fit into a single airline seat and a flight attendant tells her that she needs to purchase a second seat or exit the flight, her asshole husband tells her that she should've lost weight a long time ago and encourages her to drive to Colorado instead of flying with him and one of their female "friends." Of course, the movie later reveals that Mike has been cheating on Sheila with the female friend, who remains on the flight with him, so booting her off the plane just deepens her humiliation and his cruelty.

Many fat fliers find that the judgment of other passengers is a major cause of their apprehensiveness. "I love to fly, and my only stress in relation to flying comes from my interaction with other people," one interviewee says in *Flying While Fat*. "It's like I have a hyperawareness of my body at all times that other people don't have to think about. They don't have to think about their space and how much or how little they're taking up. I'm always trying not to burden someone else with my body."

Airlines have done little to reduce this anxiety for their plus-size customers. They take the view that fatness can be controlled. It isn't a permanent state of being. So, rather than making seats on airplanes, amusement-park rides, and public transportation bigger, fat people are encouraged to lose weight instead. Making flying as uncomfortable as possible for plus-size people is another way of punishing us for being fat. Fatness comes with a punitive burden. Whether it's being unable to find clothes that fit well, encountering doctors who treat fatness instead of illnesses, or having to shell out extra cash for an additional airline seat, fatness is treated like a crime. But we shouldn't have to apologize or make amends for existing in our bodies as they are. We also can't and shouldn't have to hole up in our homes until we lose weight, so in this no-win conundrum, how can we take back our power and assert our unequivocal right to belong?

Internet groups have emerged as a lifeline for fat people

trying to envision and forge a future without fatphobia. One example is the Traveling While Plus Size private Facebook group, where nearly sixty thousand people share tips and hacks for fellow fat people trying to make traveling less anxiety-inducing. The group was initially created by influential plus-size blogger Chastity Garner for that specific purpose, but over time it has become a safe haven where people share photos of themselves wearing swimsuits for the first time or falling in love and getting married. It's a supportive community where fat people rely on one another to build confidence and resilience in the face of an unchanging travel industry that treats our bodies as transgressive and unruly.

Yes, the group inspires, but there's also an undercurrent of sadness about how we're treated. Traveling While Plus Size is full of horror stories about fat people being discriminated against by airline employees, being unable to fit in the bathroom stalls at concerts, and being turned away from amusement-park rides because of their size. When I returned to Denver after grad school, I was excited about going to Elitch Gardens, the city's biggest amusement park, because the park had just unveiled a two-seater Batman-themed ride that pulled passengers up about thirty feet and swung them through the air. It was supposed to feel like flying. After waiting in line for an hour, we plopped down in the small black seats and pulled the restraints over our heads and toward our laps, where they were supposed to lock between our legs. My cousin's restraint snapped into place instantly. Mine refused. I pulled it down once, twice, and then a third time, but it wouldn't budge past my breasts. Before she or anybody else could notice that I couldn't fit, I said, "I changed my mind. I'm too afraid to ride this. I'm going to get a snow cone instead."

If she noticed, she didn't say anything. She just smiled at me and said she'd meet me at the food court after she got off the ride. I felt humiliated. Tears threatened to roll down my face, but I held them back because I didn't want to appear even more vulnerable or, worse, invite pity or unwelcome questions from strangers. By the time my cousin was off the ride, I had tucked away those tears and created a lie: The snow cone made my stomach hurt. I'd had too much sugar and I needed to go home and lie down. We left soon after, and I didn't return to the amusement park for another three or four years. I just pretended this kind of fun no longer existed. I put the memory in a black box, never to be revisited or remembered.

In lieu of a world willing to accommodate us, we create impromptu survival guides that help us avoid the indignities often imposed on us. AllGo, an app created by Rebecca Alexander, is an important tool in our survival kit. As we get fatter, we often have to ask ourselves questions before we leave the house for any kind of outing: Is there a weight limit? Is there parking near the entrance? Can I fit in any of the seats? Are they comfortable? If I think there's a possibility of embarrassing myself, then I'll simply stay home, making the world smaller and nixing social interaction.

AllGo tries to eliminate that uncertainty by letting fat people review places in their neighborhood with a focus on whether they accommodate fat people. It asks customers four simple questions that make all the difference: How many types of seating are available? Do most of the chairs have arms?

Do most of the benches move? Is the patio seating plus-size-friendly? Many people never have to consider these kinds of questions, but they're an essential lifeline for those of us who have felt boxed out of public spaces. AllGo's attempts to gather and publicize this basic information feels revolutionary—which only points to how far we have to go in creating a truly accepting culture.

"YOU'RE GETTING THIN, GAL!" MY MOM SAID AS I STRUTTED THROUGH THE archway of my parents' front door. "Thanks, Mom," I muttered. It was the fifth time I'd heard the compliment in recent weeks. Close friends, other relatives, and even my then live-in partner were showering me with incessant praise about the sudden slimming of my wide waist and rotund thighs. Before this unexpected weight loss, I weighed more than I ever had before, a direct result of a crippling depression during grad school.

I had always been a larger woman, accustomed to squeezing into a size 22. With large breasts that sprouted at age nine and stretch marks that followed not far behind, I'd never existed in a body other than the large one I walked through the world in. As one of several larger women in my immediate family, confidence in my body was instilled in me. Diet culture had occasionally weighed on me, but, for most of my life, losing weight had never been my objective.

Yet, the weight fell off when I finished graduate school, moved to a metropolis, and fell in love with a man who loved to hike. I was eating less because I wasn't using food to comfort myself. My weekends were suddenly filled with hiking and trips. Soon, I was much lighter, and I realized that all of their praise indicated that I was being treated better because I now took up less space in the world. Nobody sighed when they had to sit beside me on a plane. Waiters didn't make snide comments when I ordered a burger instead of a salad. I could walk into a store and search for clothes instead of pretending that I was only looking for accessories because none of the clothing fit. It felt relieving to walk through the world without worrying, at all, about my body. I wasn't concerned about people snickering as I walked by or men calling me a "fat bitch" when I rejected their advances. I fit into seats now. I fit in. I fit.

Losing weight impressed upon me how much discrimination women of size face. Unlike racial, gender, and sexual orientation discrimination, weight discrimination has yet to be defined by the legal system. There is no mention of it in the United States Constitution, in Title VII of the Civil Rights Act of 1964, the Rehabilitation Act of 1973, or the Americans with Disabilities Act of 1990, legislation that is supposed to prevent discrimination in the workforce. A 2008 study conducted by the Rudd Center for Food Policy & Health at Yale University concluded that weight discrimination is as prominent as racial discrimination. Once people reach a body mass index (BMI) of 27, which puts them in the "overweight" category, they face increased discrimination in the workforce, with 43 percent of those surveyed reporting that they've experienced weight bias from coworkers, supervisors, and employers.

According to Rebecca Puhl, director of research at the Rudd Center, weight discrimination has increased by 66 percent in the past decade. Puhl told the publication *INSIGHT Into Diversity* that "compared with other forms of discrimination, it's the third most common for women, the fourth most common for men." From Puhl's perspective, the cause is much more straightforward than people realize. It boils down to words. As a society, we have adopted "fat talk"—seemingly benign comments such as "Do I look fat in this?"—into our language, thereby promoting the stigmatization of overweight people, particularly women.

When you've been fat for as long as I have, backhanded compliments aren't only common but expected. Most people, regardless of their size, ethnicity, or gender, have experienced some form of a backhanded compliment from coworkers, friends, and even relatives, though it can take some time—and paying close attention to language—to recognize them. I first began noticing backhanded compliments when I would come to visit family during college breaks and summer vacations. The women in my family, including my mother and grandmother, would look me up and down before embracing me and planting kisses on one or both cheeks.

"You look cute, but I don't know why you wore that on the plane," they'd say. Or, after I made the decision to chop off all my hair in 2011, I'd hear, "You're lucky you have the face to pull that off because, *chile*, I don't know if I could've done it." I'd smile and laugh when they made these snide comments, mostly because I didn't know how to respond. Were they insulting me? Were they complimenting me? Or were they doing both without realizing it? Even now, I find it difficult to decipher if I'm being complimented or insulted.

Delving deeper into a backhanded compliment reveals its true nature: it's designed to insult someone while also disarming that person's defenses. Instead of registering the offensiveness of the overall comment, we focus on what we consider to be positive or affirming.

In a study conducted at the Harvard Business School, researchers Ovul Sezer, Alison Wood Brooks, and Michael I. Norton studied how backhanded compliments operate in the workplace; for anyone who has dealt with passive-aggressive bosses who are seemingly never satisfied, the findings aren't all that surprising. Backhanded compliments serve two goals in office settings: eliciting liking or respect while also conveying superior status.

"We propose that people believe that delivering the 'compliment' part of a backhanded compliment will garner the benefits of flattery for liking, while using the 'backhanded' part to avoid being seen as lower status," the researchers concluded. "With backhanded compliments, flatterers specifically place recipients lower . . . because flatterers both control the comparison set and in fact exclude themselves from that set." Though backhanded compliments are ineffective and fail to achieve the two goals determined by the researchers, there is a perceived payoff for the flatterer, who believes their words reinforce their position. "Implying that the recipient is of low ability may harm the recipients' perceptions of their own competence, decreasing their motivation—likely making the flatterer look better by comparison."

There are a slew of backhanded compliments that are often directed toward fat people, and most end with the age-old "for a big girl" modifier. We've all heard the classic "You're so pretty for a big girl," but there's also "You dress so nice for a big girl! Your clothes really flatter your figure." The one comment that I despise in every way is "Wow, you're so confident!" It implies that our confidence is undeserved and therefore illegitimate. Actress and overall Hollywood titan Mindy Kaling eloquently spoke to this issue when she appeared on the cover of *Parade* in May 2013. "I always get asked, 'Where do you get your confidence?' I think people are well meaning, but it's pretty insulting," she said. "Because what it means to me is, 'You, Mindy Kaling, have all the trappings of a very marginalized person. You're not skinny, you're not white, you're a woman. Why on earth would you feel like you're worth anything?'" All of these insults are couched in flattery but are really designed to make us feel less-than, unworthy, and unwanted.

Fat women aren't supposed to be confident in a system that persuades us to hate our bodies. We're constantly told, in subtle and not-so-subtle ways, that our bodies are a result of gross negligence. We've used sweets and butters to destroy our bodies, so our obligation should be to chisel away at the fat, not get more comfortable in our skin. We are expected to be defined and imprisoned by our bodies, so how dare we exude confidence? How dare we choose clothing that flaunts what others perceive as imperfections? How dare we opt to live a full life instead of one limited by our size? A fat woman's confidence shouldn't be considered an act of resistance, but alas, in a world where our bodies are policed, especially through language, it becomes one.

In a 2014 study published in the *Journal of Health Psychology*, researchers followed around fifty overweight and obese women, noting every fat-shaming comment or outright insult that people around them made either behind their backs or directly to their faces. On average, these women were the targets of three fat-shaming incidents per day, ranging from insensitive teenagers making mooing sounds behind their backs when they walked past, to a dentist complaining a patient might break his chair. A more recent study, reported on by *The New York Times* in a December 2017 article titled "Fat Bias Starts Early and Takes a Serious Toll," examined fat-shaming in social media. Two of the most common words to appear after "fat" in fat-shaming tweets are "girl" and "lady." Many times, this fat-shaming was targeted at women. Confident fat women are not outliers, but we deal with so much bullshit. We're told how to dress, what to eat, and even what should offend us. This slew of weight bias persists because it's considered an acceptable form of discrimination. "There seems to be a public perception that the stigma against obesity is justifiable because it provides motivation for people to lose weight," Puhl explains. "But what we find is that the opposite is true. The social stigma causes binge eating, exercise avoidance, and other behaviors that lead to increased obesity."

Weight discrimination is as serious and widespread an issue as "obesity" itself. Proponents of weight discrimination legislation believe that including obesity in the Americans with Disabilities Act could be a viable solution. This protection would guarantee plus-size workers employment and provide them with the right to sue if they felt that they were discriminated against in the hiring process based on their weight. When filling out an application, we're able to check our gender, ethnic background, and educational status—but there's no way to self-disclose if we are among the category of people whose body size makes us highly vulnerable to discrimination. The question is: Should there be? Most supporters of laws against weight discrimination, including the International Size Acceptance Association, believe so.

Most of the solutions are plausible, reasonable, and give fat people the same access to the American dream that was promised to us all. As a curvaceous woman with huge ambitions, I am determined to bend the rules of the corporate world, because a larger shirt and pants size should not be the ultimate factor in deciding whether I should be promoted to CEO. Since being diagnosed with heart failure and losing a significant amount of weight, I now fit in an airplane seat without accommodations. I can get on a roller coaster without wondering if I'll be forced to exit because I can't fit in the seat. But I remember vividly the look of pity a flight attendant offered me as she passed me a seat belt extender on a flight to New Orleans. I remember the man sitting next to me who snickered and muttered to himself that I should "lose weight," further contributing to the humiliation. I, of course, eloquently told him to fuck off, but couldn't hide the shame I felt. Instead of addressing the institutional problem, the rules force the individual into guilt for daring to be larger. People of size, including me, deserve to be accommodated and treated with respect.

IMAGINE THIS: A CHILD, WHO HAPPENS TO BE FAT, BEGINS SCHOOL FOR THE first time. Their body is visibly larger than those of their classmates, but nobody blinks an eye. Their classmates approach them with the same level of kindness, crankiness, sass, or excitement that they bring to anybody else. When that child breaks out their lunch box—or purchases lunch from the school—nobody surveils what's on their plate or comments about how much they eat or how quickly they eat it. They're able to play freely on the playground without anybody staring at their body as it moves, and there's no size-focused bullying or taunting. In gym class, exercises are adapted for their body's needs and abilities, and they're never singled out to do extra laps or pull themselves up on a rope to prove their fitness. The school nurse doesn't weigh them and pass the number on the scale on to their parents and their doctors or whisper about that child's BMI during parent-teacher conferences. In fact, school nurses aren't weighing children at all.

During social studies, when it's time for that child to learn history, the curriculum includes pioneers such as Steve Post, a radio personality who organized the first-ever fat-in in New York City's Central Park to "protest discrimination against fat people." These activists' efforts to make the world more equitable for fat people are considered worthy of study and discussion. And when that child leaves school at the end of each day, their parents don't ask them about what they've eaten. Instead, when they arrive home, they eat a snack (or not) and turn on the television. There are no ads about dieting, and their favorite shows have a number of fat characters who aren't on a continuous quest to lose weight and don't exist simply as punch lines. When that child goes to the doctor, their BMI isn't constantly put forward as a topic for conversation, and every children's clothing store they venture into has plenty of clothes in their size.

Fatness is a part of their life, but fatphobia—at school, at home, in what they consume in pop culture and how they shop—is not. This child will come into their own without the internalized belief that their body is defective. They won't have to spend their adolescence, or even the rest of their life, attempting to correct themselves and fit into an "ideal" that never should've existed to begin with. Imagine a society full of free children who spend time exploring the magic of their existence rather than picking themselves apart.

I often struggle to believe that this free world—a liberated future—could ever come to be. Fatphobia, racism, sexism, homophobia, transphobia, classism, and other forms of oppression are a matrix that we're taught to see ourselves and the world through. But abolitionist organizer Mariame Kaba has taught us that hope is a discipline, and creating a new world requires us to imagine what's beyond our immediate periphery. There's another future on the horizon, if we can just make it to the edge of the earth to see it, feel it, usher it into existence. Don't fat children deserve to live in that world? Don't we all? When it comes to ending fatphobia, we're closer than we've ever been before, but we're still so far away. How can we survive? We have beautiful clothes. We have models and magazine covers and swimsuits. These improvements grant fat people more dignity, which is something we all deserve, but that doesn't account for the rampant fatphobia in other aspects of our lives. Only a single state, Michigan, prohibits weight-based work discrimination. Trans people of all sizes are being assaulted at every level of government. Black people are still being subjected to racism. Fat children are still being taught they're unworthy. Doctors are still discriminating against us. We're not much closer to a free world than we were before body positivity became a popular hashtag on Instagram. We don't have time to waste; we must return body positivity to its fat liberation roots.

That will take far more focused thought from people from all backgrounds, but here's where we can start: Lobby the Federal Communications Commission to permanently suspend all dieting-focused commercials. Create regulations for the representation of fat people in network television. Bankrupt Weight Watchers and similar programs, or at least support a campaign that stops their program from targeting children. Fight with medical schools to ensure that antifat bias is included in the rigorous training doctors are subjected to. End police violence. End police altogether. Push for legislation that protects trans people and ends workplace discrimination against them. Fight for universal health care and basic minimum income. Fight. Fight. Fight. I have been fat nearly my entire life. I will be fat forever. That doesn't mean I—or any of us—have to live in a fatphobic world. In fact, it's more feasible and healthy to change our fatphobic world than attempt to control fat people's bodies. We can create something new; we deserve something new. We deserve to survive. We deserve to live. We deserve to be fat in plain sight.

ACKNOWLEDGMENTS

I never thought I'd be able to finish this book, but the fact that I am writing these acknowledgments means that I have. Finally. *Finally.* Hallelujah. It has taken four years; a publisher who didn't understand the vision; a returned advance; two chronic illnesses; and a relentless, determined agent; but every failure was worth it. This story, my story, won't sit on shelves, gaining dust because it hasn't been read, and for that, I have so many people to thank. God is my guiding light. I am only here because of God's grace, which has given me a renewed appreciation for every breath, every minute, every day. Every day above ground is a blessing (word to Beyoncé), and I don't take any of it for granted.

My agent, Sarah Phair, stuck beside me through every iteration of this book, holding me to deadlines (my Kryptonite) and pushing to get me the editor and the publisher this manuscript deserved. I am forever grateful. I could not have asked for more from Sara Birmingham, my editor, and Ecco, my publisher, who treated me with so much attention and care. Sara's guiding hand brought this collection together, and she pushed me to make it the best collection it could be. I will forever appreciate those come-to-Jesus Zoom meetings about putting less pressure on myself and getting the book done. Sara was also relentless in securing the best possible cover, going through artist after artist until Vivian Rowe knocked it out of the park. Sara, I thank you. Vivian, it is an honor to have your art on the cover of this book. Thank you for understanding the vision and bringing it to life. I owe so much to every person at Ecco and HarperCollins who touched this book, including the copy editors who caught all my mistakes, small and large, and helped shape this manuscript. Thank you.

My family continues to anchor me to everything that matters in life: faith, peace, calmness, humility, and sanity. My parents, Dwain and Kelly, housed me when I became ill, changed their diets to accommodate my restrictions, and accompanied me to every appointment, procedure, and test. I can't ever repay you for nursing me back to health, but I will attempt to for the rest of my days. My grandma, who always reminds me that I'm not famous enough not to call her, lights up my life. Thank you for always sharing your wisdom with me—and for helping to choose the final book cover. You, Mom, and Aunt Tracey (hey girl! I love you!) made the perfect choice. To my brother, Andrae, who always has my back, thank you for your unending support. Keep knocking at the doors of your dreams; one of them will eventually open. Shameika, my nearest, dearest friend, I love you beyond words. You show up multiple times in this book because you always show up for me, and I can't even express how grateful I am for your friendship. You inspire me every single day. Please don't forget it.

Friendship makes my world go round. When I'm in a rut, I call my friends. When I need advice, I call my friends. When I need to celebrate, I call my friends. They've given my life so much meaning and I am grateful for every single one of them: Antasha, Christian, Julia, Erica, Briana, Candyce, Marissa, Nneka—thank you for being my friends. And to Teré, the man who I've chosen and who has chosen me, life only gets greater from here. Thank you for your never-ending support and love and for going after your own dreams with so much vigor. Julia, those two Six of Wands cards were spot-on!

Weightless is a part of a canon of books that address our culture's obsession with thinness and attempts to reorient us around a more inclusive world that considers the impact of fatphobia on fat people. I owe a great deal of gratitude to those writers and thinkers who helped shape my own thinking and the foundation for this book. They held the door open, and it is my honor to walk through it and continue this work alongside so many other people. To all the sweet girls in my life—my nieces, Melonie and Maddison; my goddaughter, Nola; and my chosen pumpkins, Josie, Nevaeh, and Zahla, I hope this book finds you when it needs to, just as so many books I've read found me at the exact right, divine moment. This world is hard on girls, especially Black girls, but I hope you're able to navigate a world that's more tolerant, more loving, more inclusive, and more authentic than the one I inherited. And if not, I hope you're able to fight for that world, if only for yourself.